JON KESSLER

THE PALACE AT 4 A.M.

CHARTA

Design
Ohad Meromi

Editing
Harald Falckenberg and Jon Kessler

Copyediting
Sam Frank

Design Coordination
Daniela Meda

Editorial Coordination
Filomena Moscatelli

Proofreading
Charles Gute
Barbara Dalla Libera

Translations
Nicholas Grindell: pp. 69-72 (Falckenberg), pp. 51-58 (Seyfarth)
Harald Falckenberg: p. 105 (Heiss)
Harald Falckenberg and Larissa Hilbig: pp. 105-107 (Foster), pp. 112-118 (interview)

Copywriting and Press Office
Silvia Palombi Arte&Mostre, Milano

Sales Department
Antonia De Besi

US Office
Francesca Sorace

ISBN: 978-88-8158-658-5

Printed in Italy

Photo Credits
Rick Haylor: front cover, pp. 10, 23, 44, 92, 144
Isabelle Woodley: pp. 3, 4, 5, 24, 25, 73
Andrew Ohanesian: p. 6
Tom Powel: pp. 8, 9,11, 12, 14,15, 16, 17, 19, 26, 27, 29, 38, 42, 43, 121-123, 126-132
Egbert Haneke: pp. 18, 75, 76, 77, 79, 80, 81, 85-91, 93-104
Hikari Yokoyama: pp. 28, 37
Walter Remy: pp. 74, 78, 82, 83
Michael Dannenmann: pp. 133-141
Bernd Borchardt: pp. 124, 125
Magnus Magnusson: back cover
Works on pages 121-123, 126-128, 130-132 were produced in collaboration with Rachel Gladfelter at Dieu Donné, New York.
p. 68: Paul Thek, *Untitled (Sodom and Gomorrha with Hot Potatoes),* 1972
Courtesy of Sammlung Falckenberg
p. 68: Gordon Matta-Clark, *Conical Intersect*, 1975
Courtesy the Estate of Gordon Matta-Clark and David Zwirner, New York

Edizioni Charta srl
Milano
via della Moscova, 27 - 20121
Tel. +39-026598098/026598200
Fax +39-026598577
e-mail: edcharta@tin.it

Charta Books Ltd.
New York City
Tribeca Office
Tel. +1-313-406-8468
e-mail: international@chartaartbooks.it

www.chartaartbooks.it

Printed in September 2007
by Tipografia Rumor, Vicenza

This publication follows the exhibitions

Jon Kessler: The Palace at 4 A.M.

Curated by Alanna Heiss

P.S.1 Contemporary Art Center,
Long Island City, New York
October 30, 2005 – February 6, 2006

Phoenix Kulturstiftung/Sammlung
Falckenberg, Hamburg-Harburg
June 10, 2006 – December 31, 2006

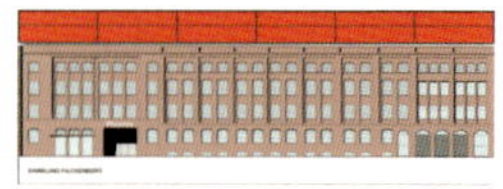

Galerie Hans Mayer,
Düsseldorf
July 28, 2007 – August 16, 2007

This publication was made possible also for the support of:

Deitch Projects, New York

Galerie Hans Mayer, Dusseldorf

Phoenix Kulturstiftung/Sammlung
Falckenberg, Hamburg-Harburg

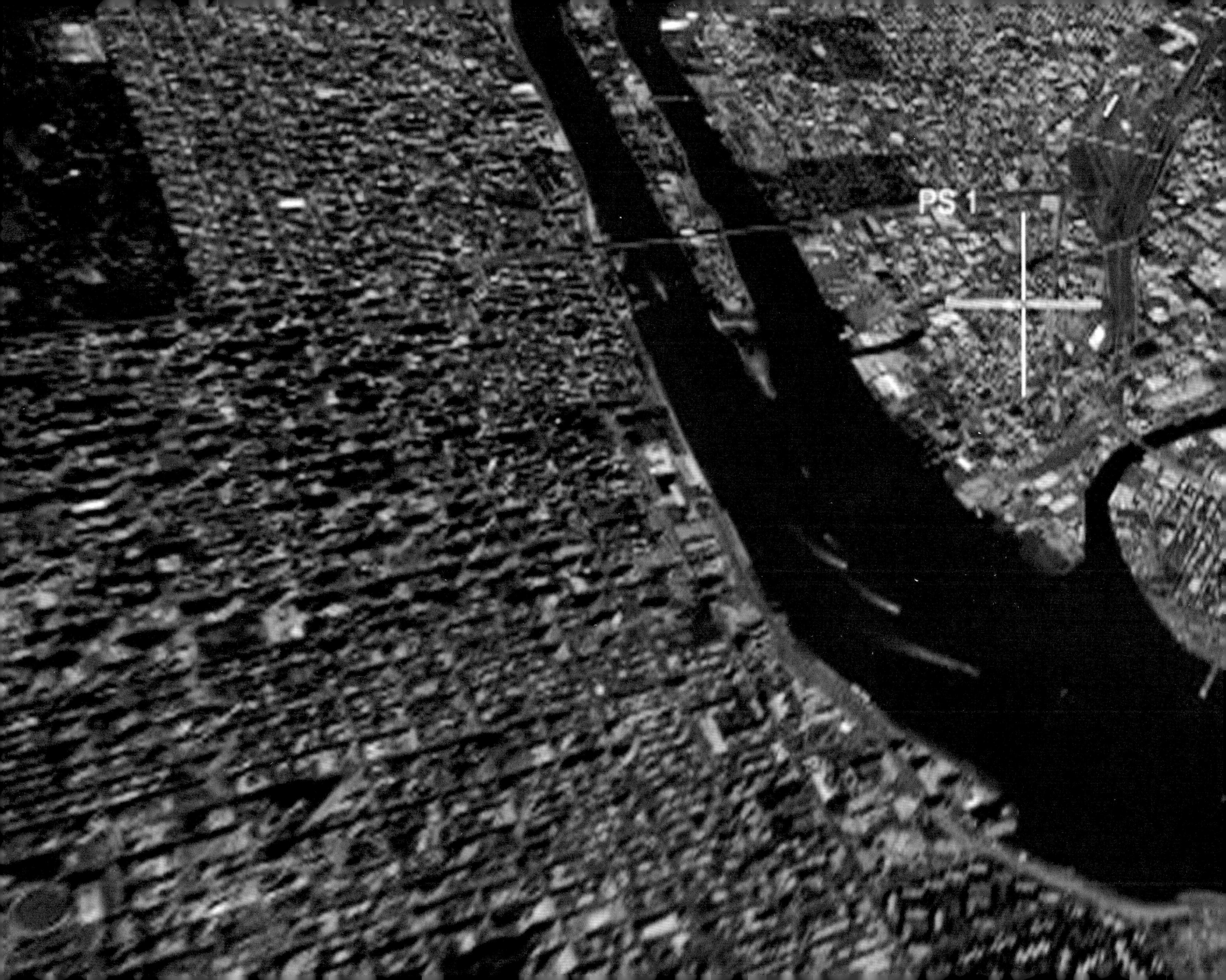
PS 1

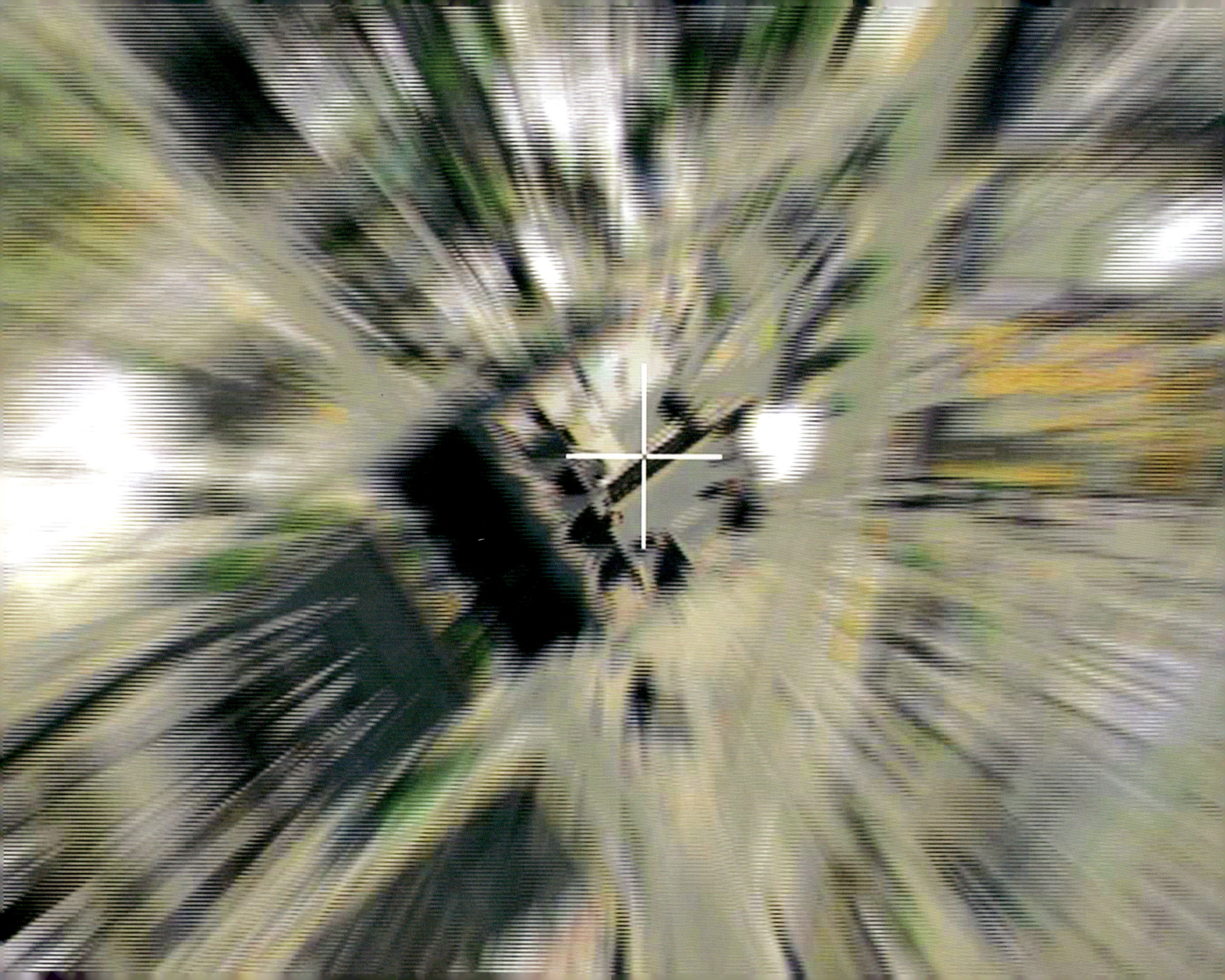

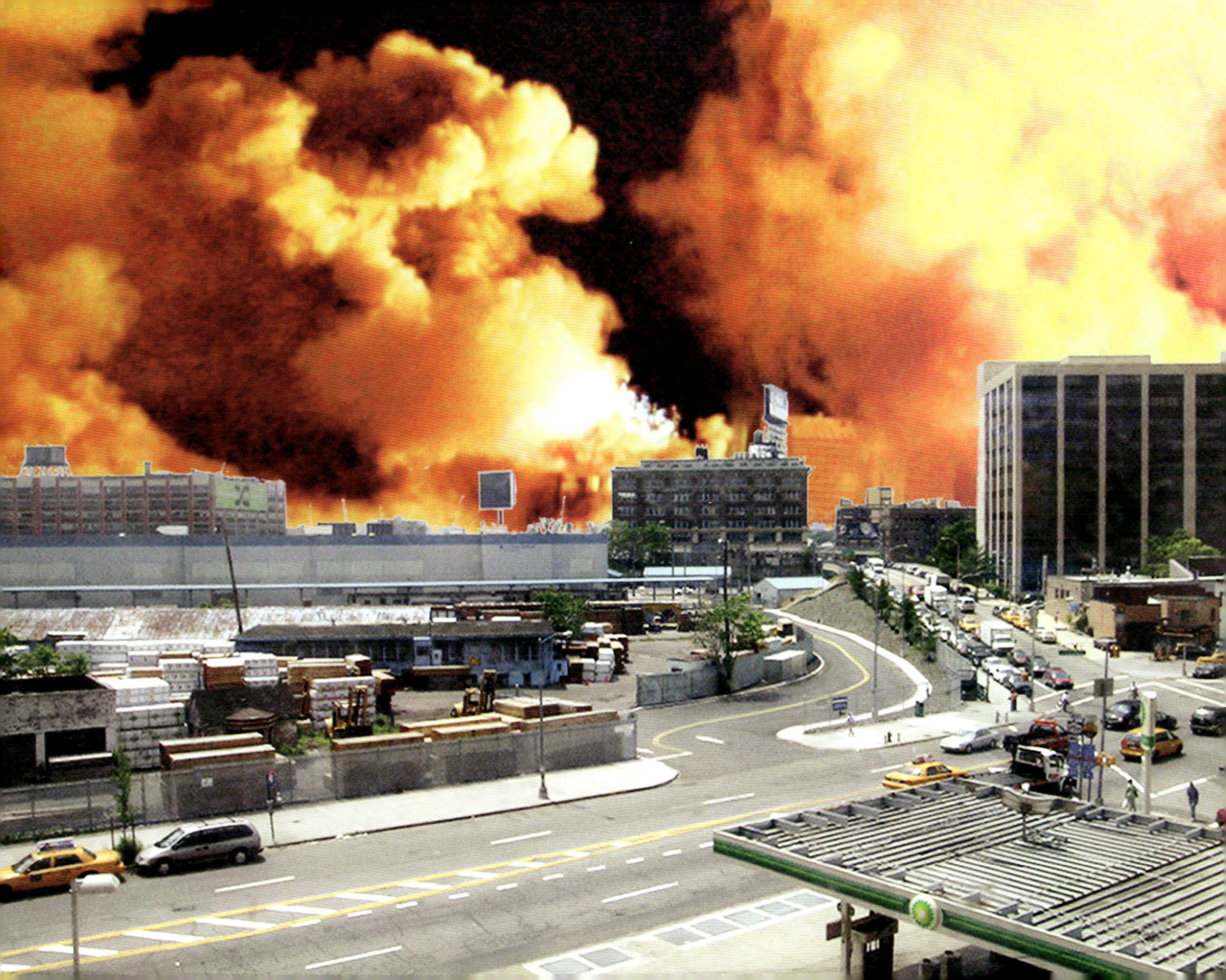

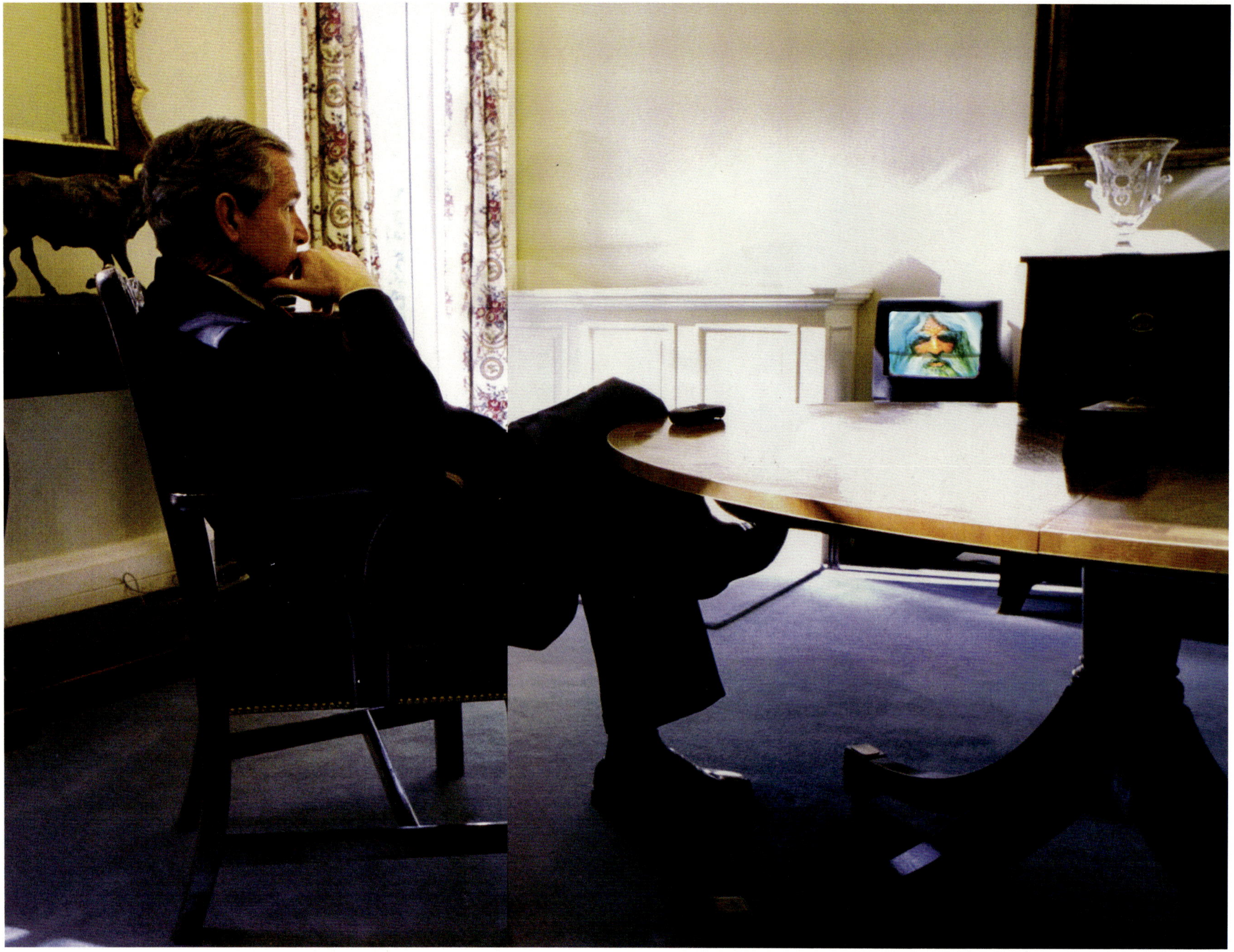

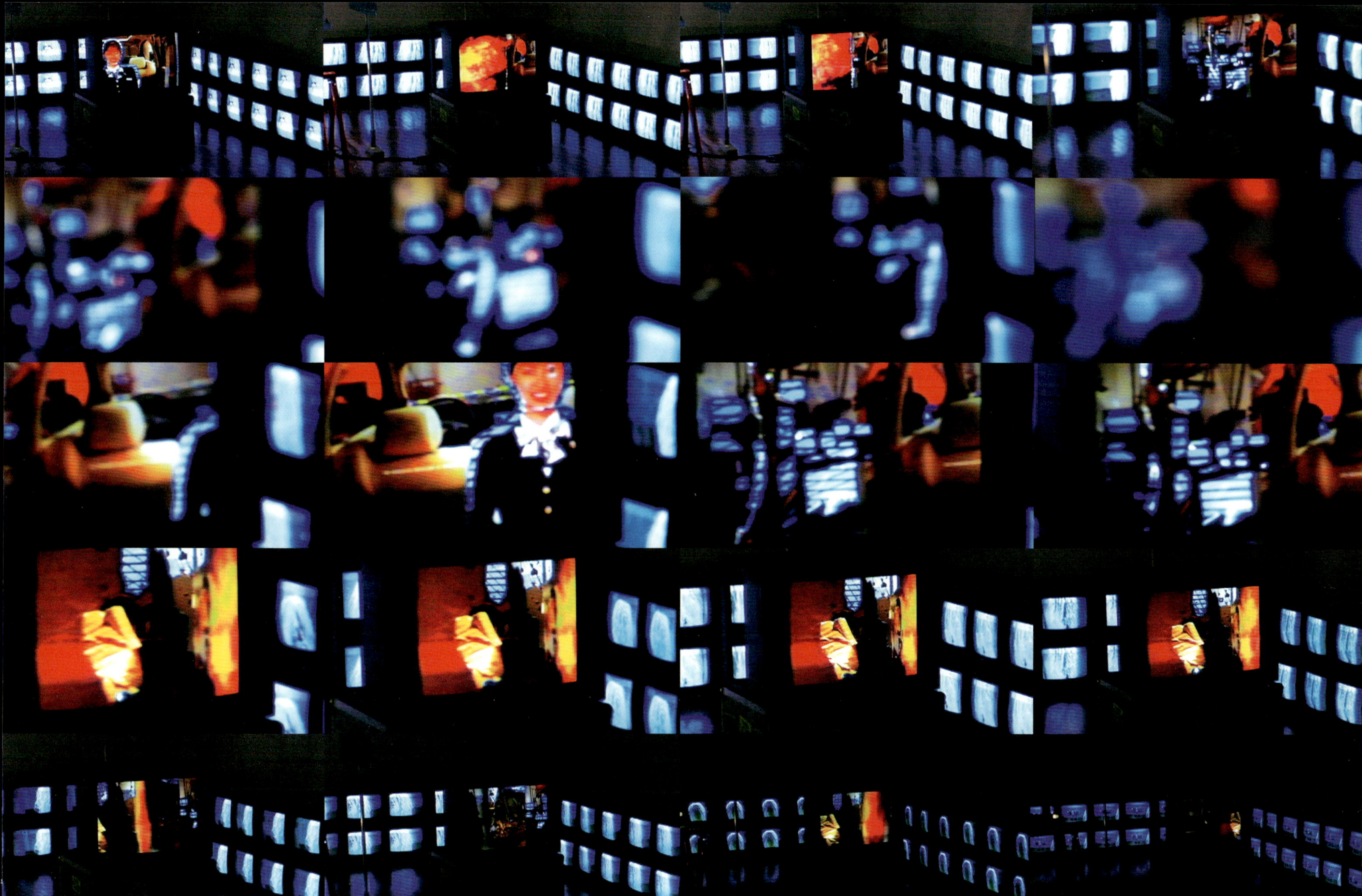

New York

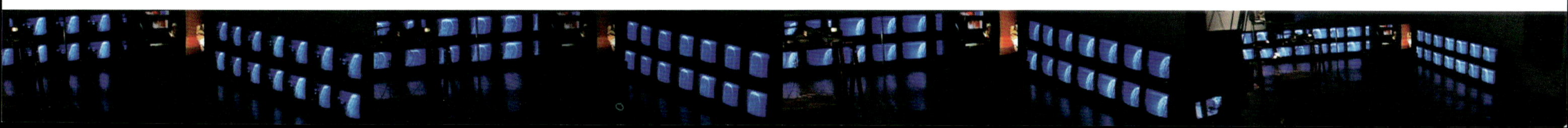

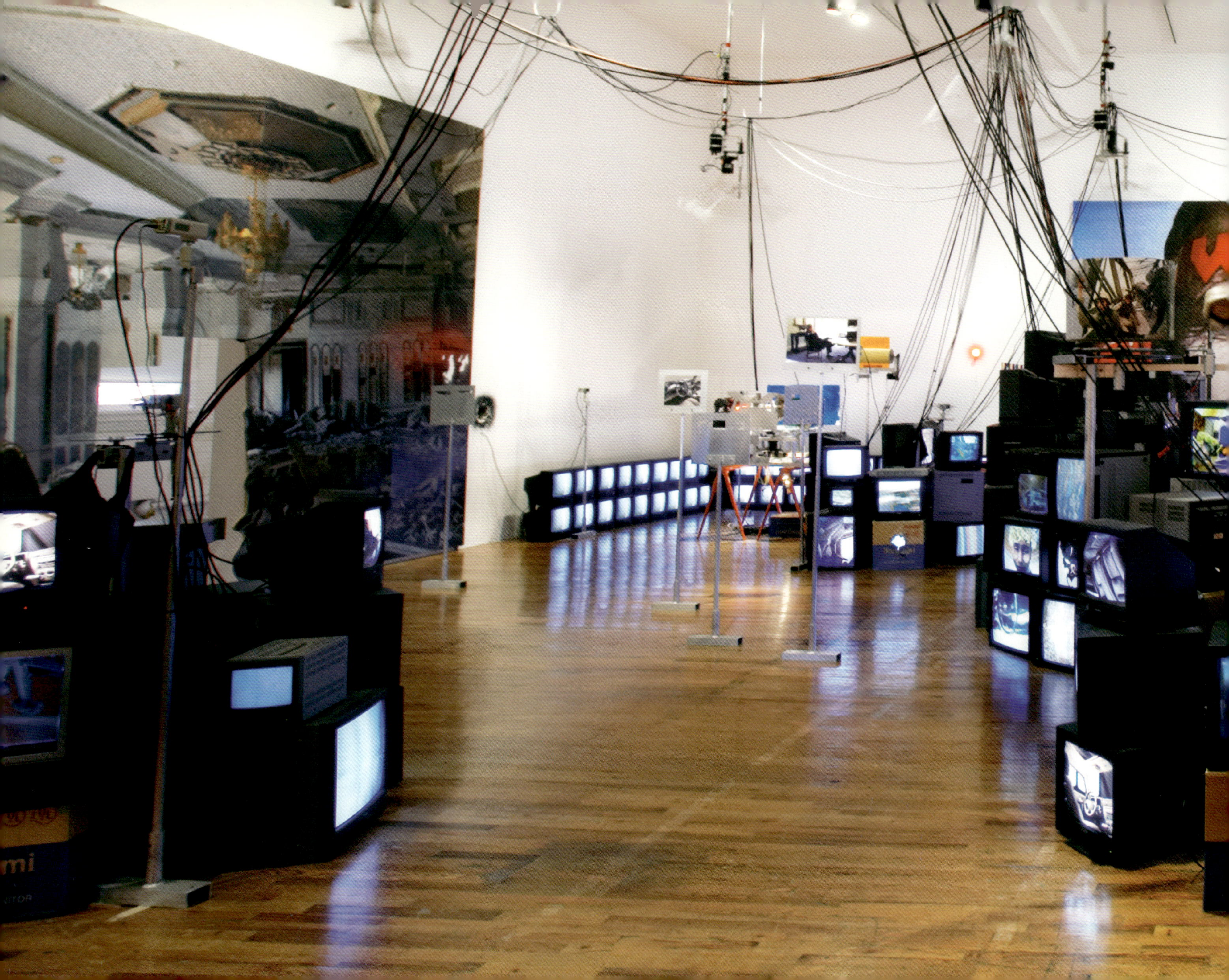

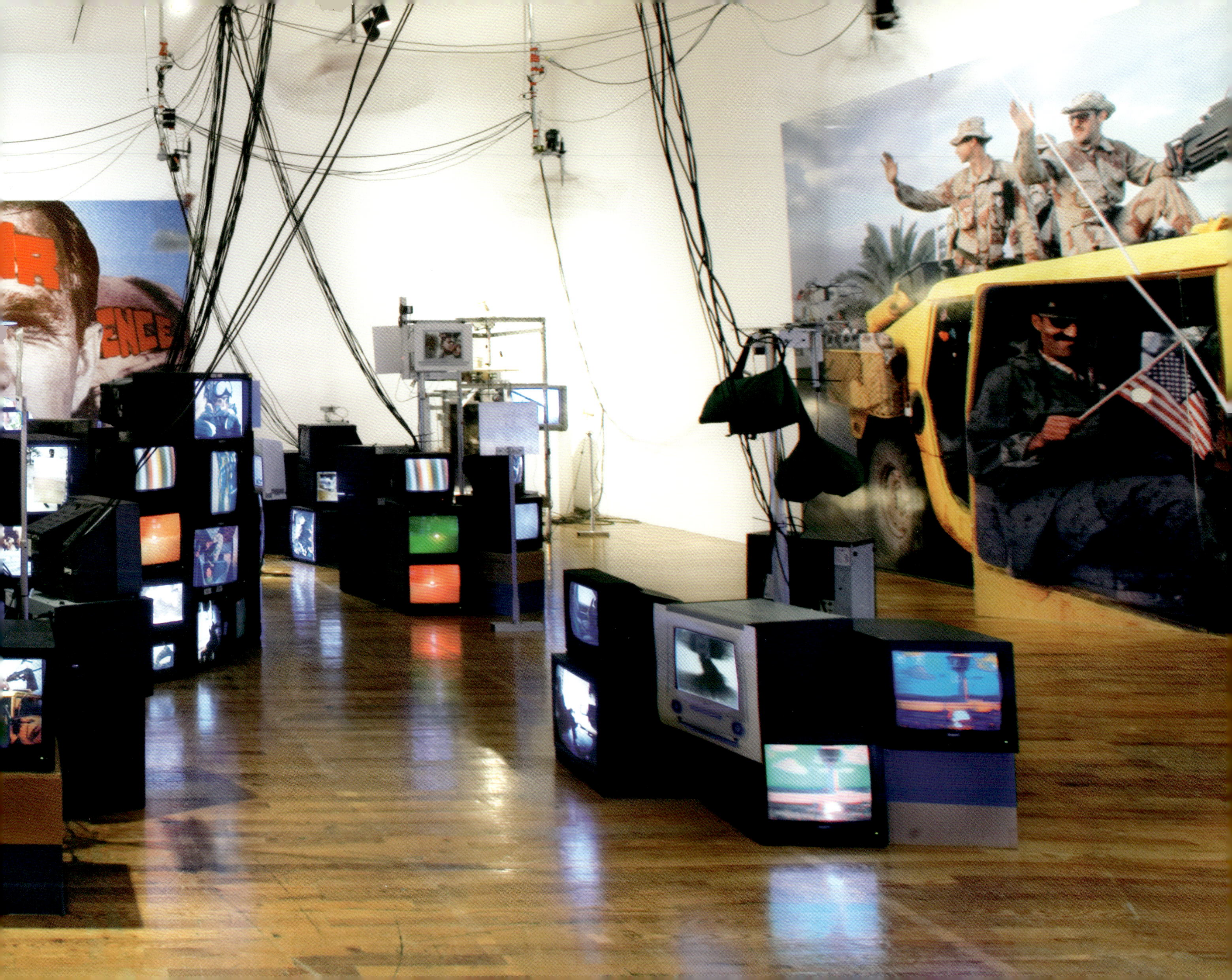

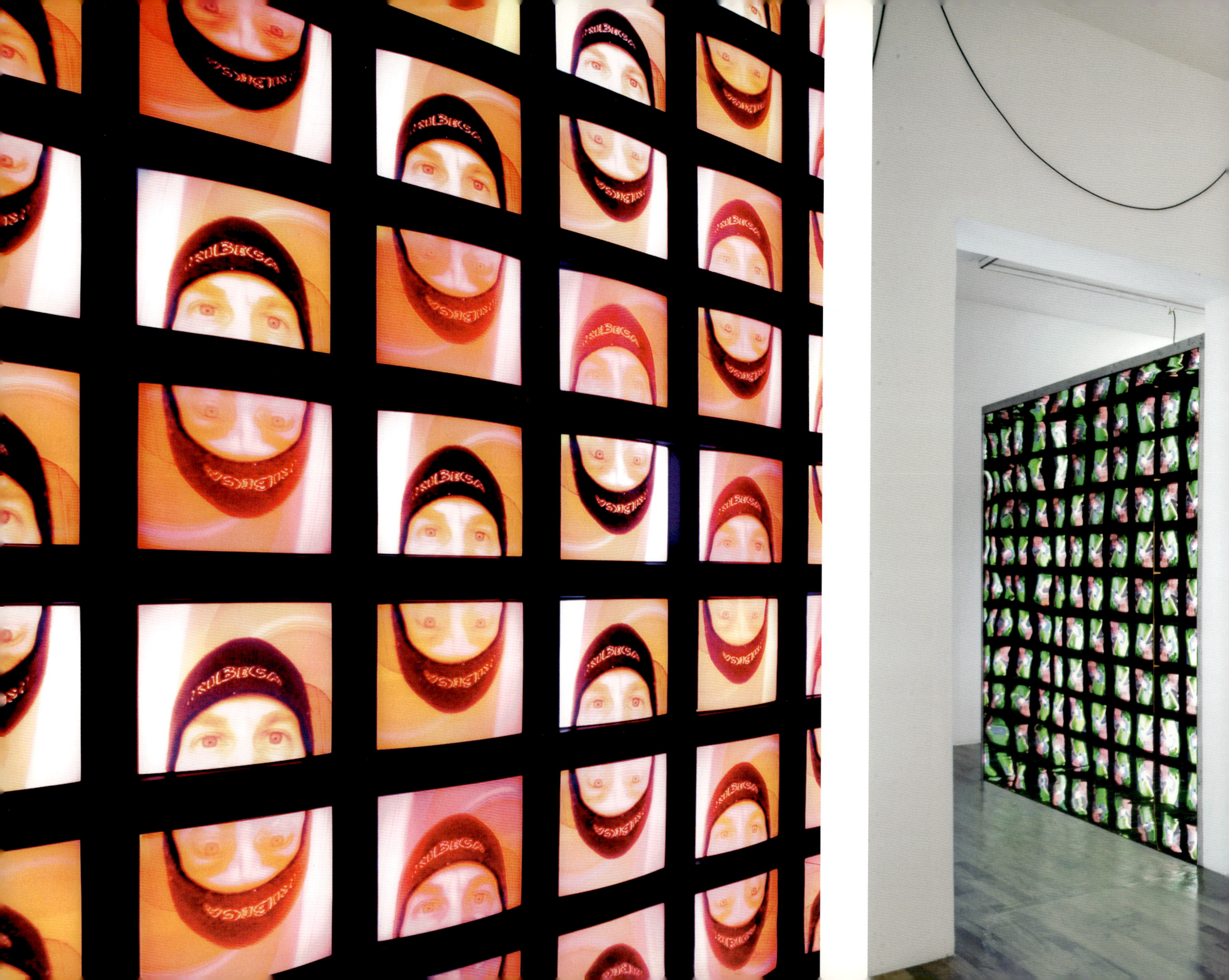

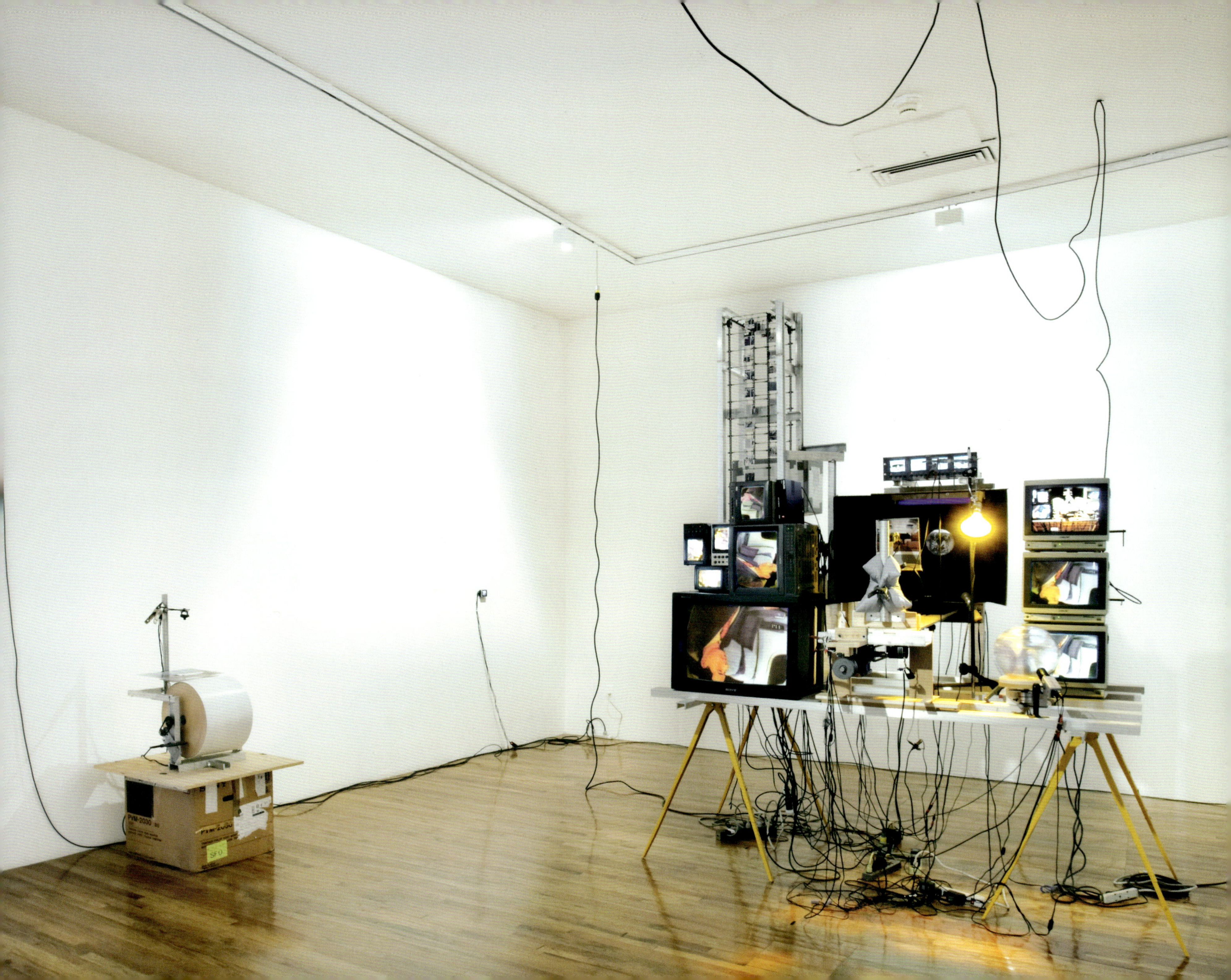
PVM-2030
SFO

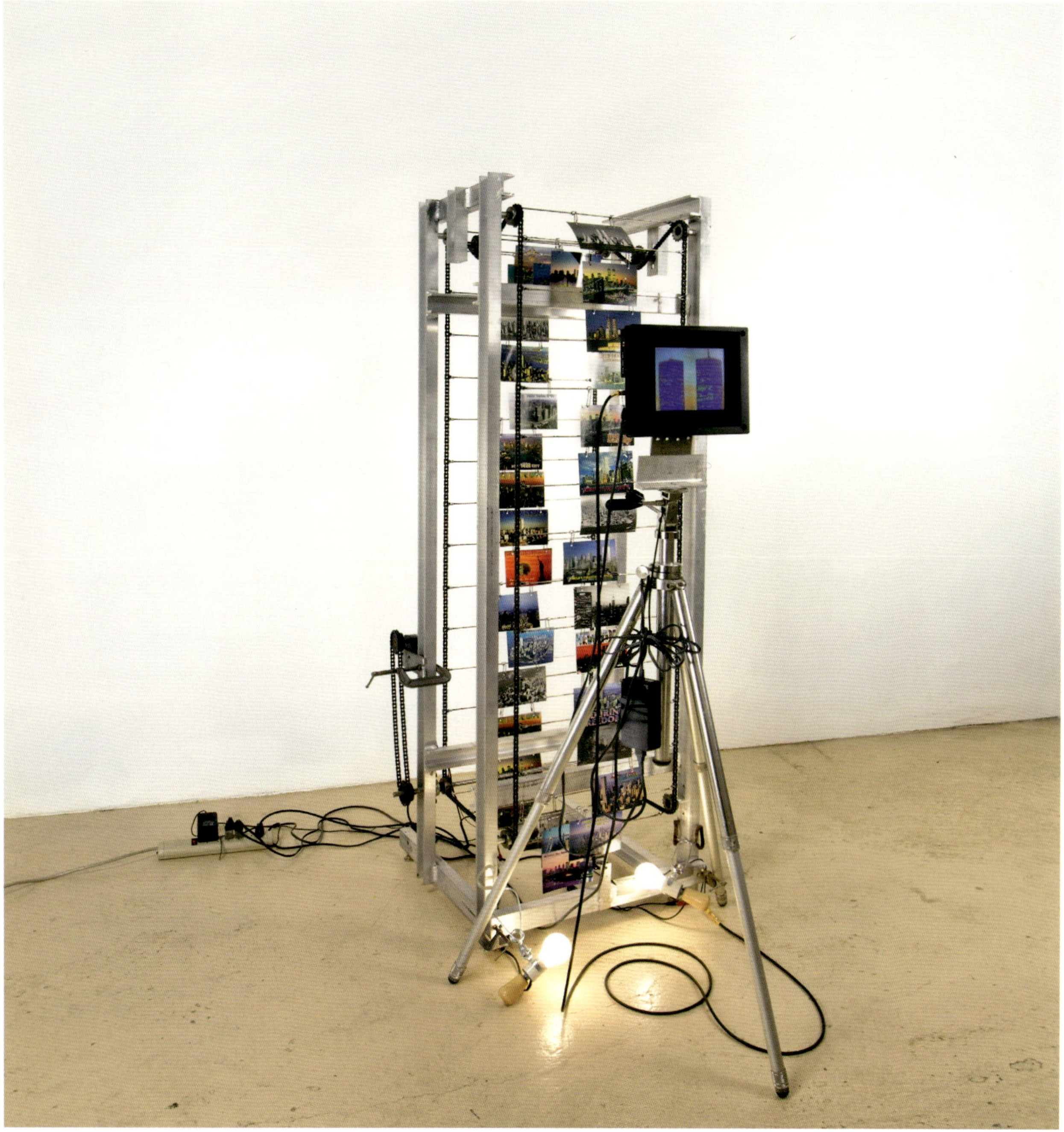

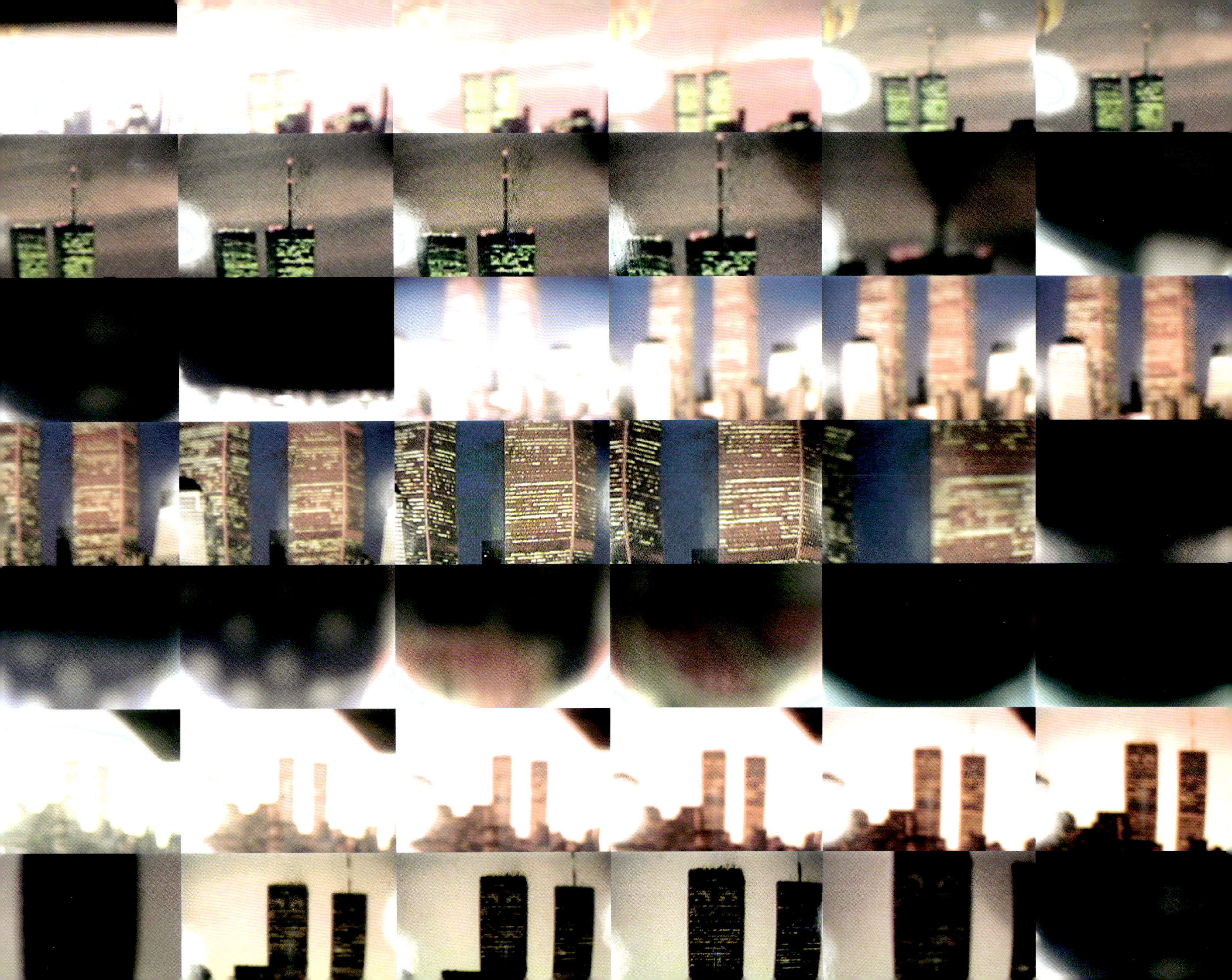

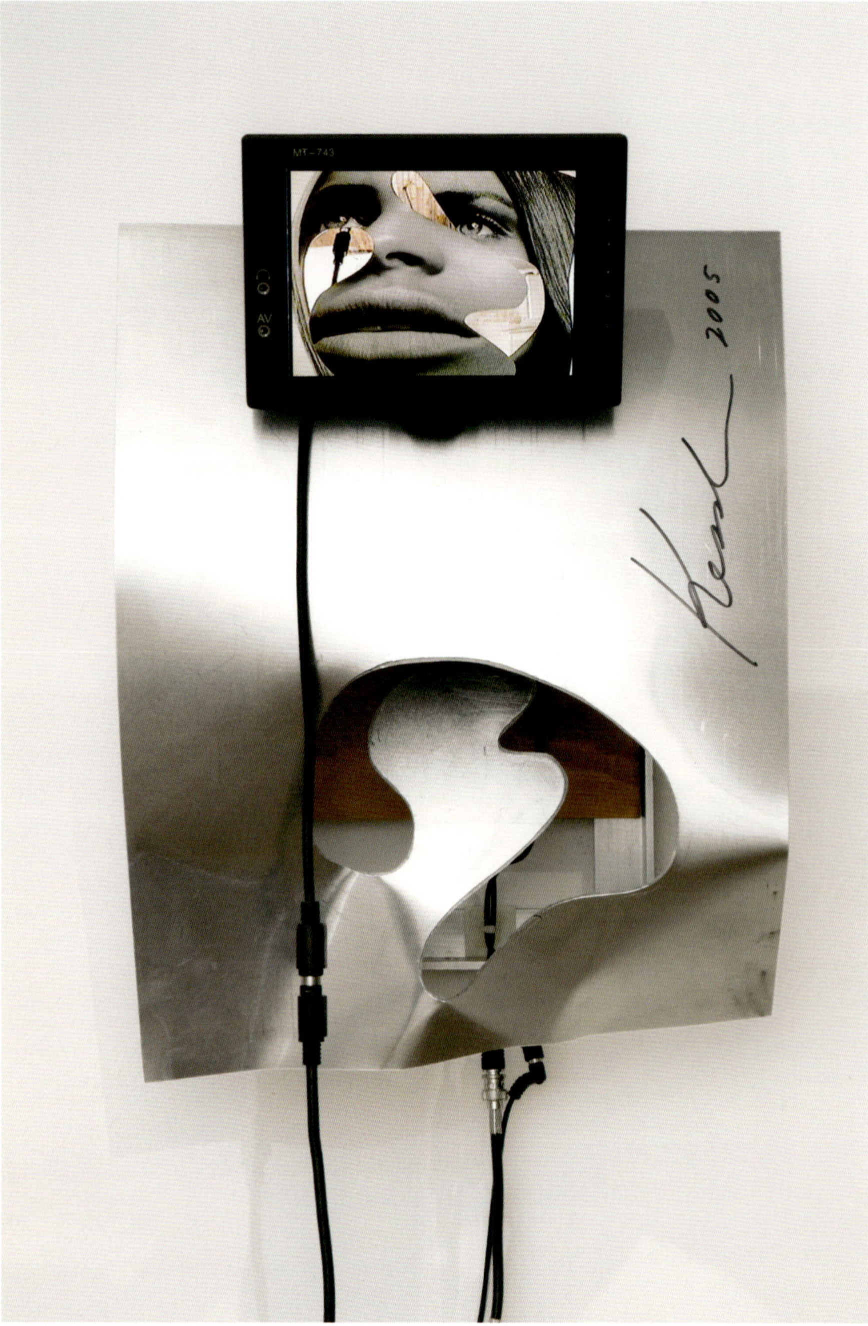
AV
2005

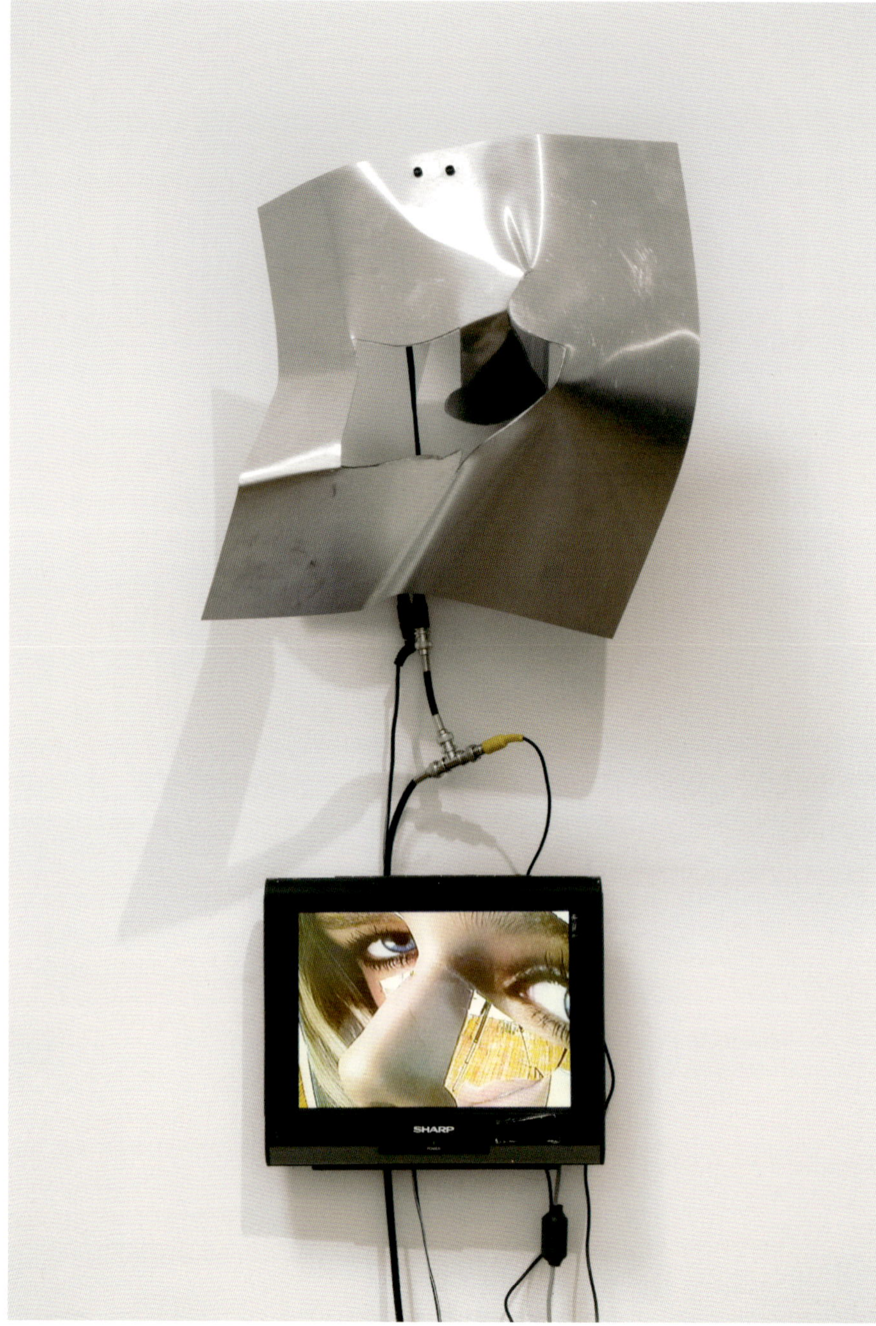
SHARP

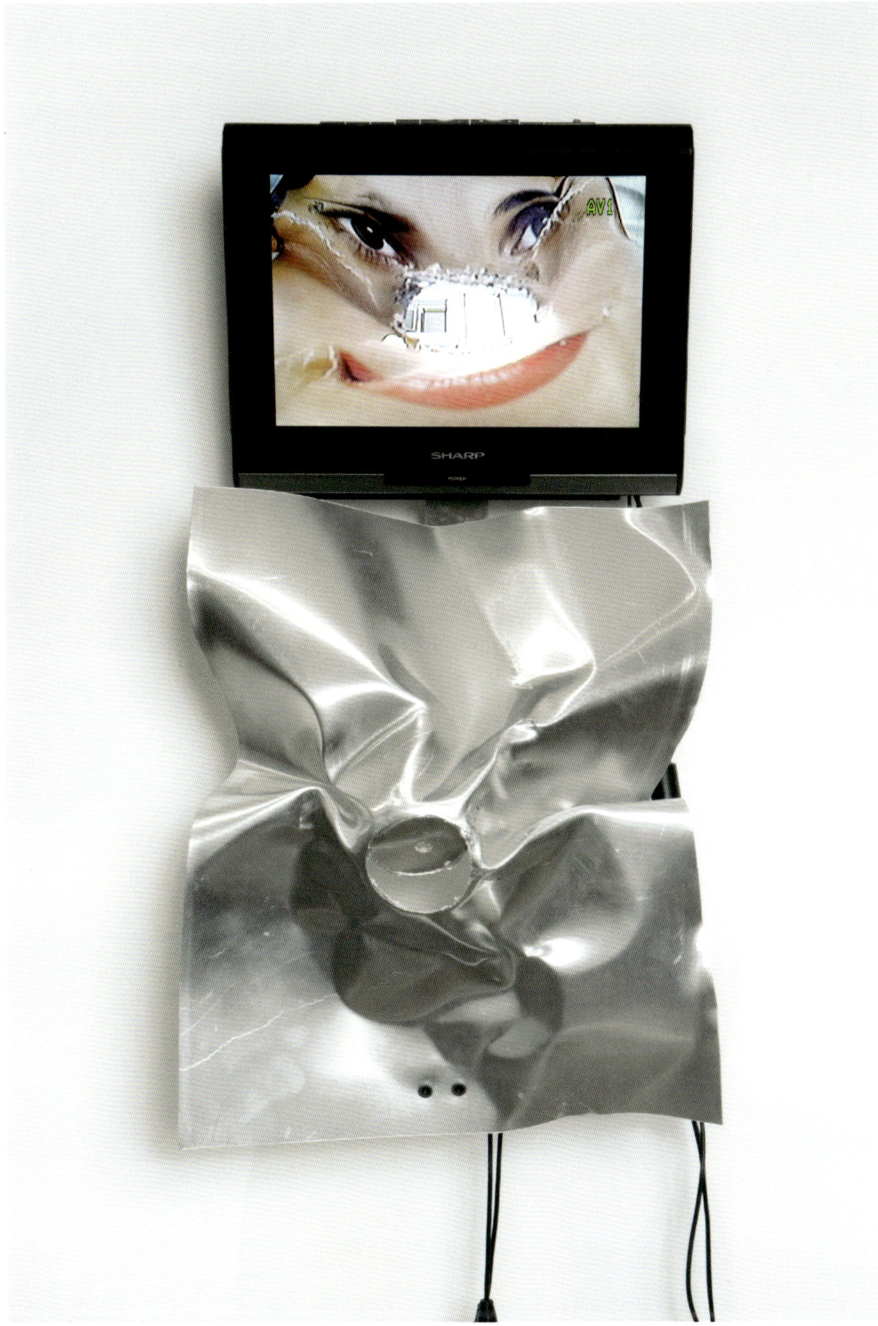
AV1
SHARP

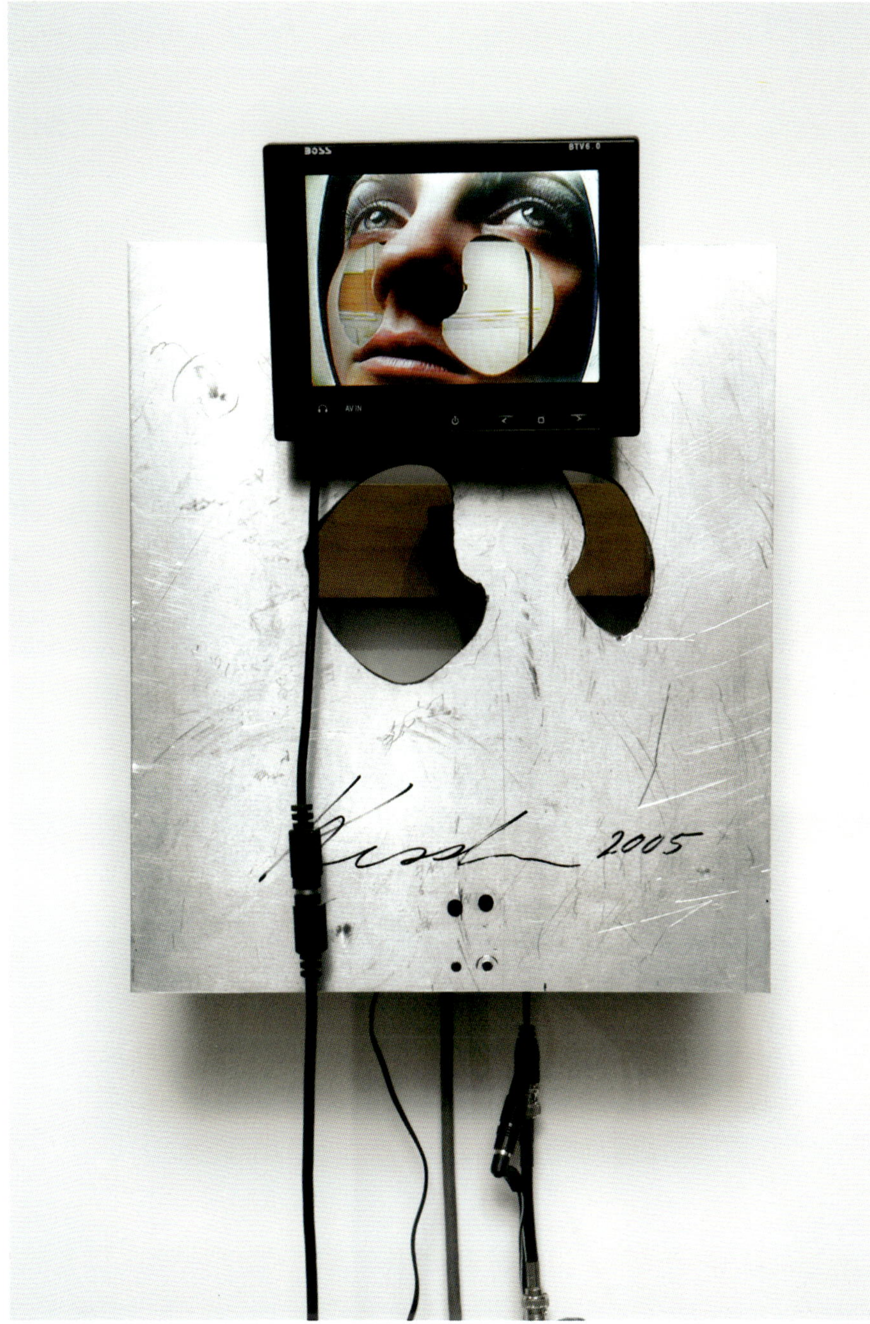
BOSS
2005

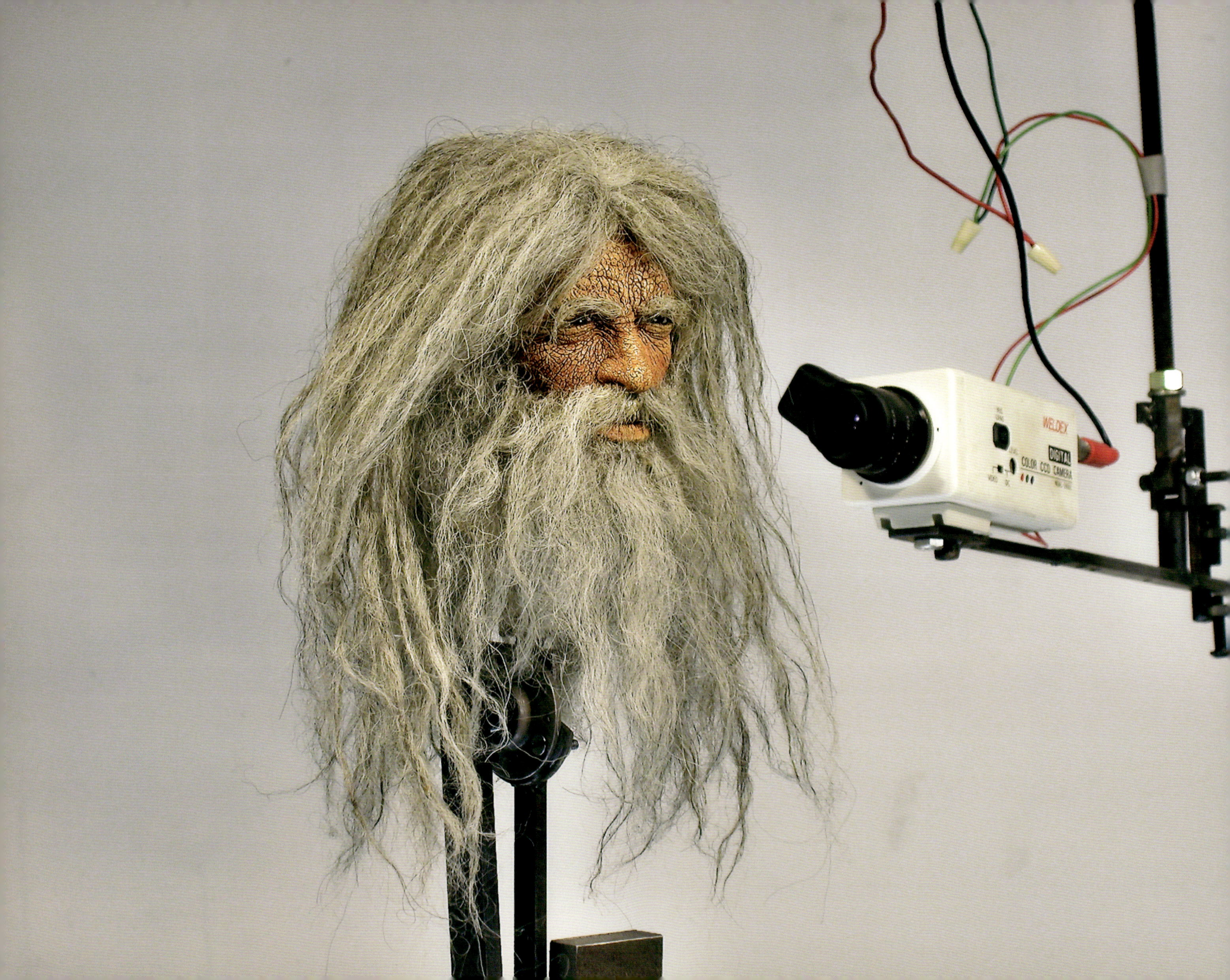
WELDEX
DIGITAL
COLOR CCD CAMERA

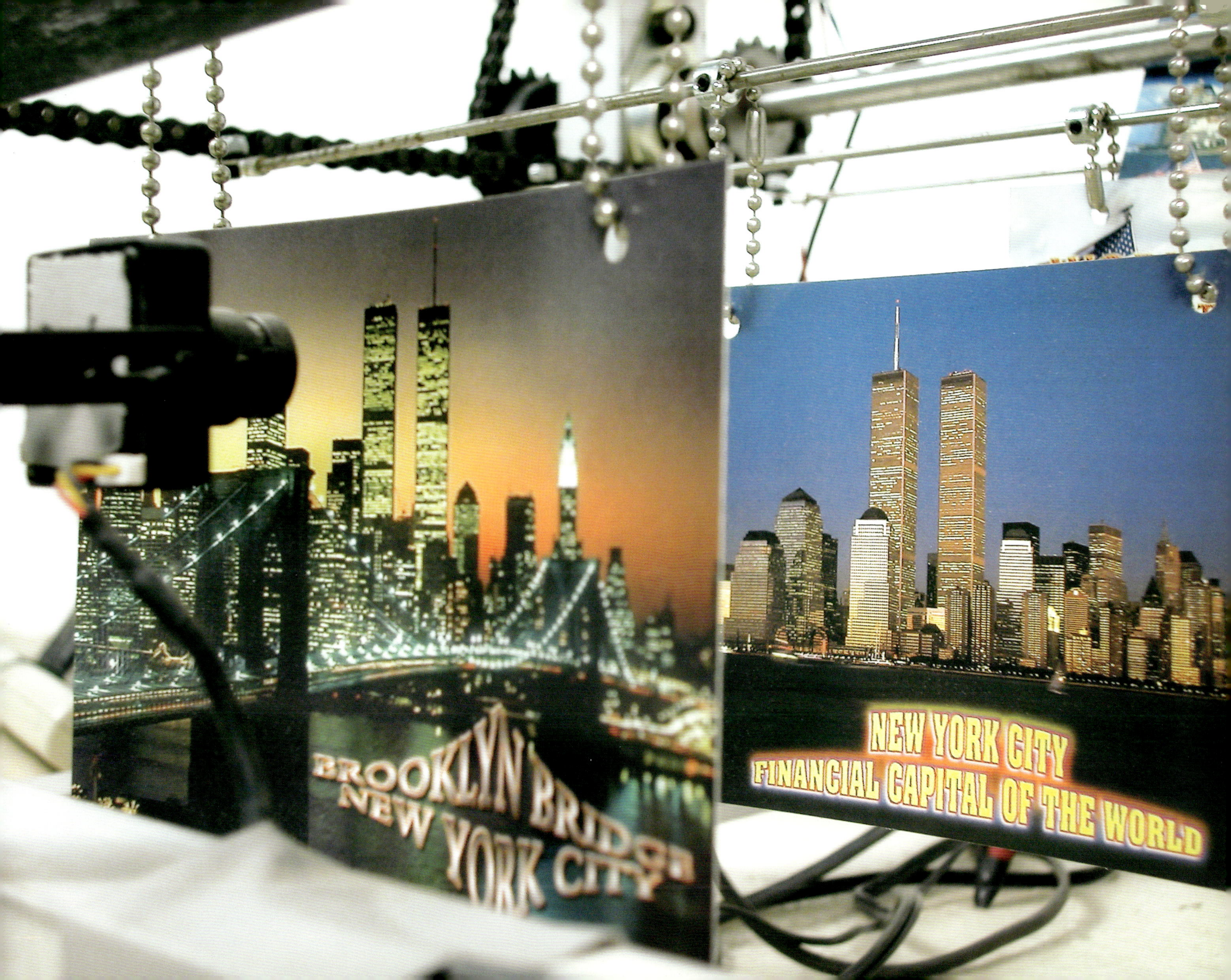
BROOKLYN BRIDGE
NEW YORK CITY
NEW YORK CITY
FINANCIAL CAPITAL OF THE WORLD

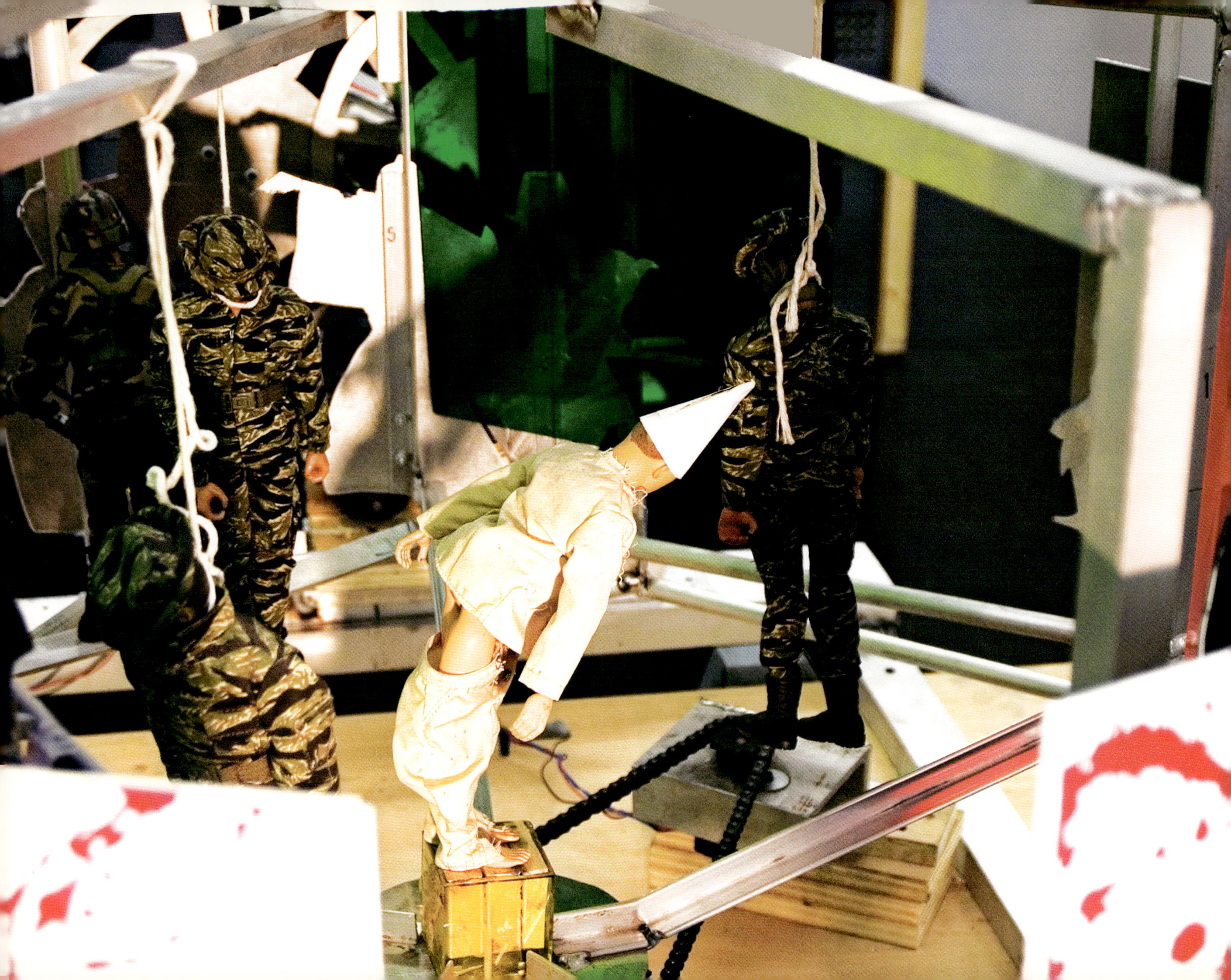

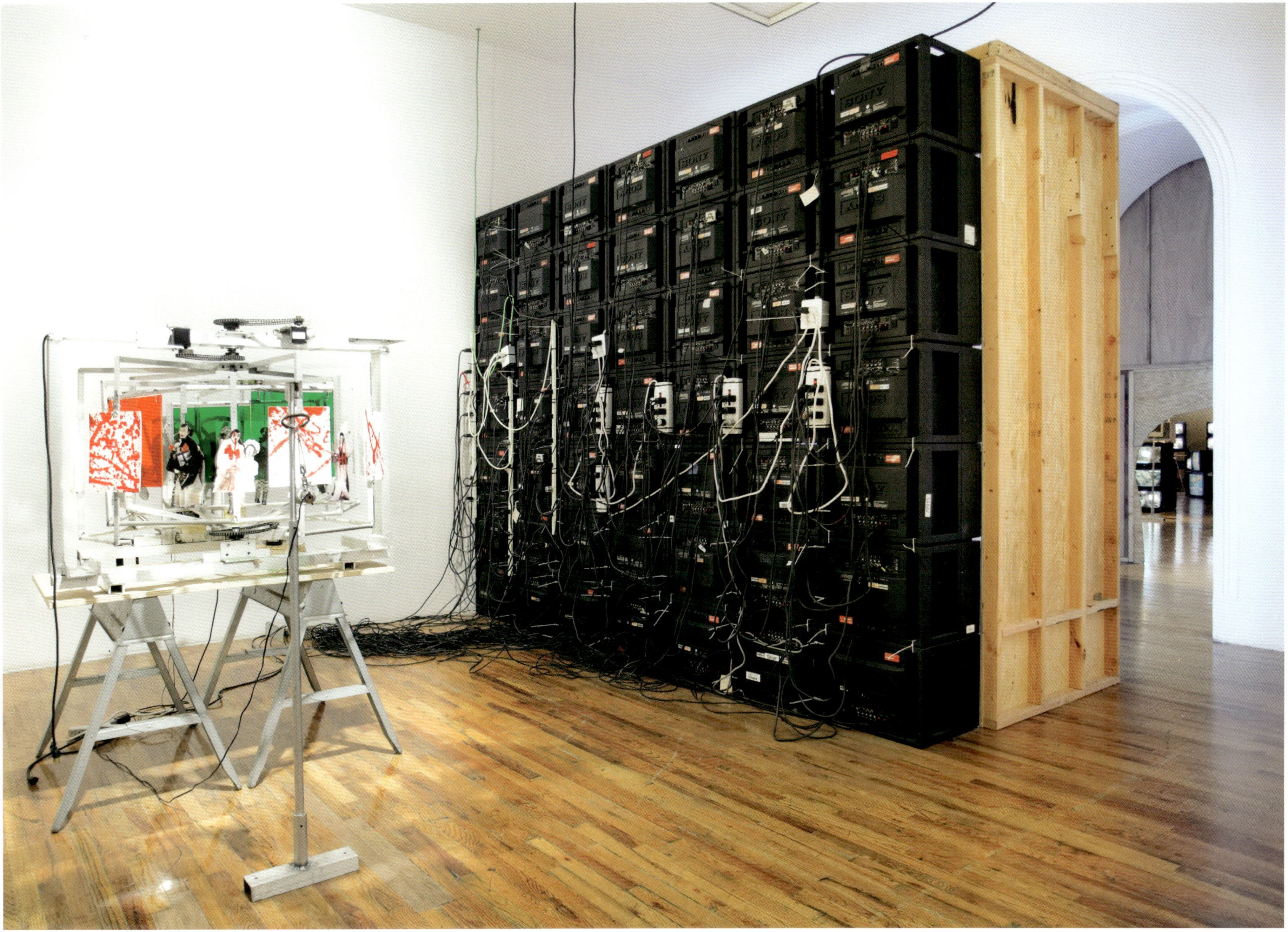

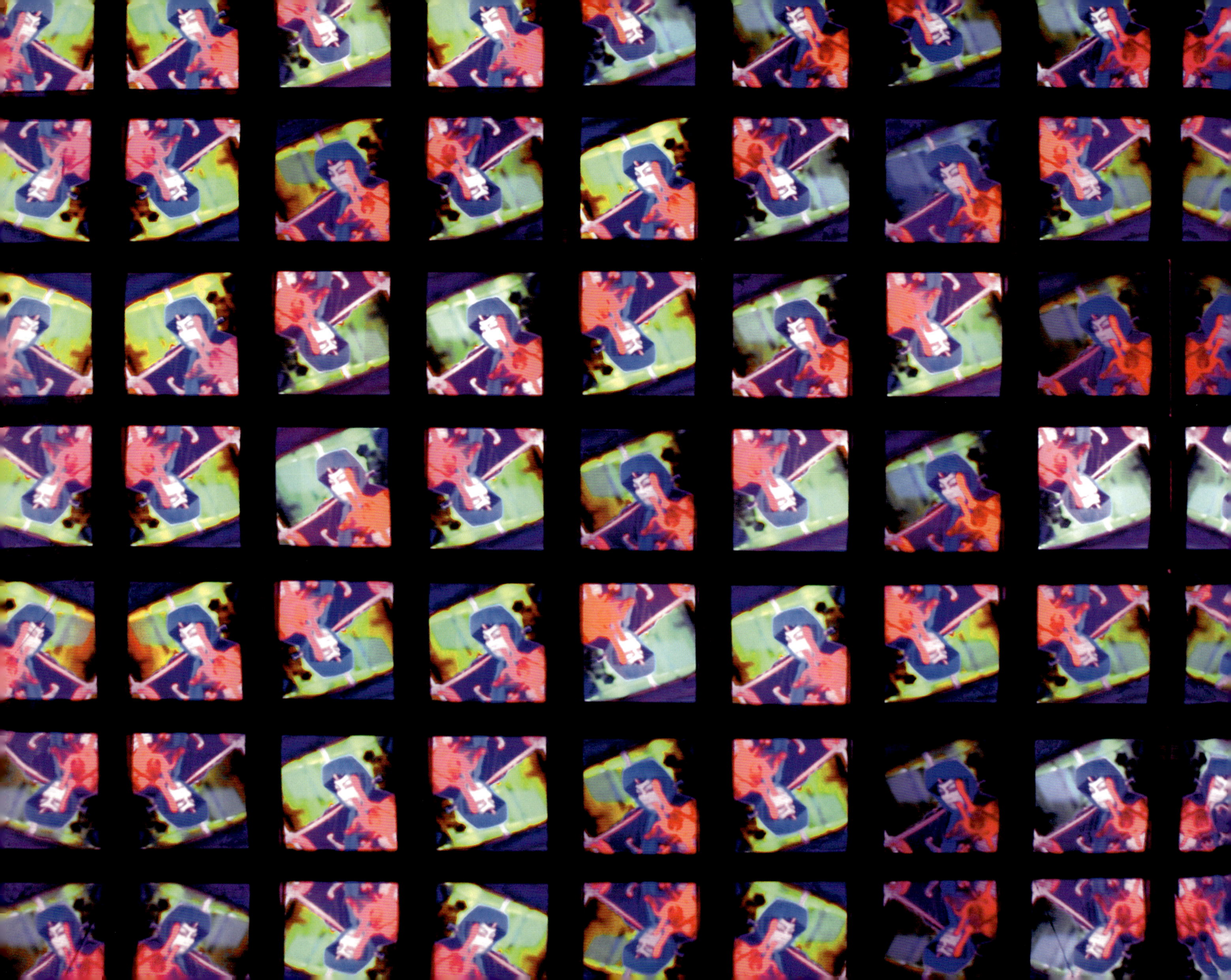

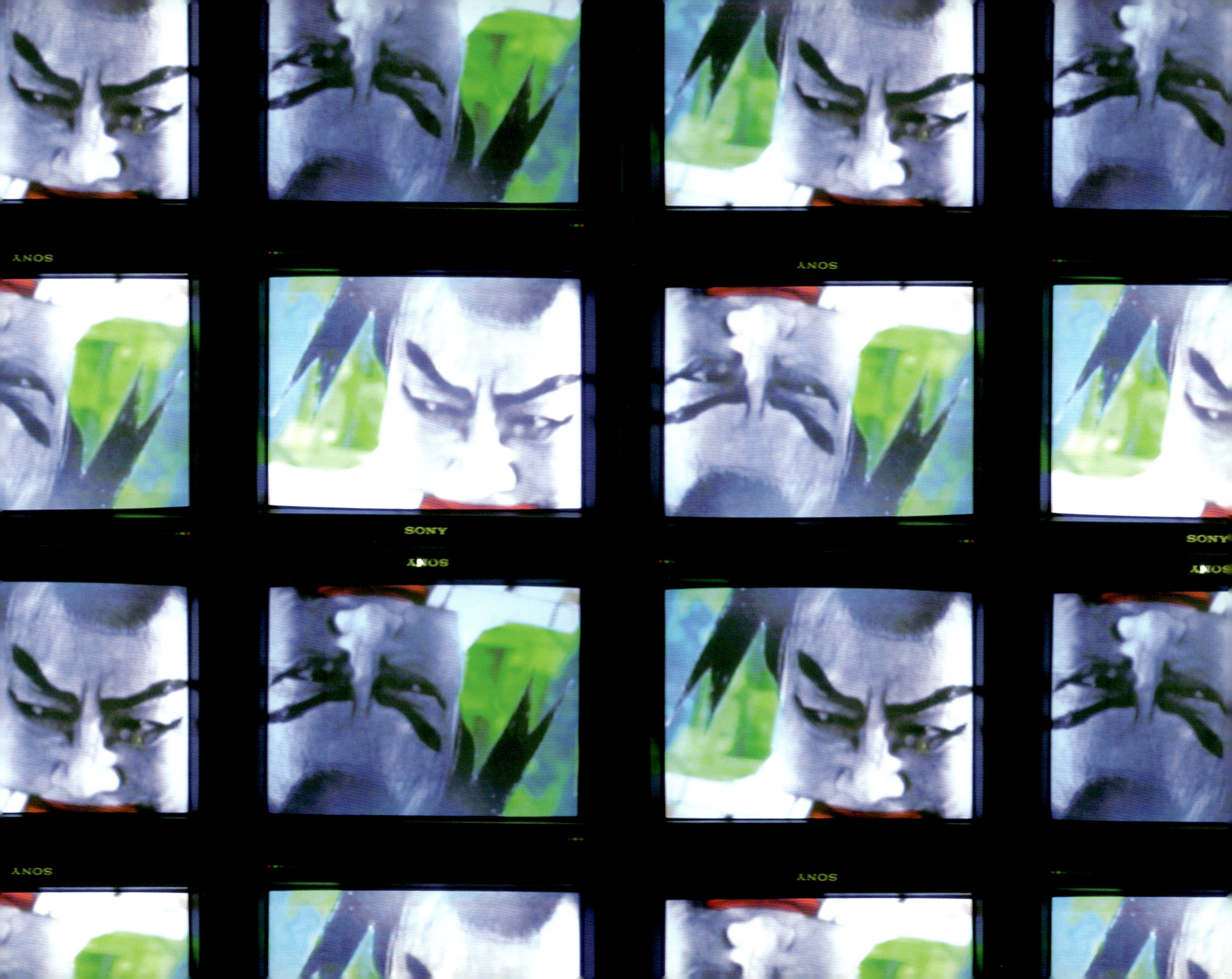
SONY
SONY

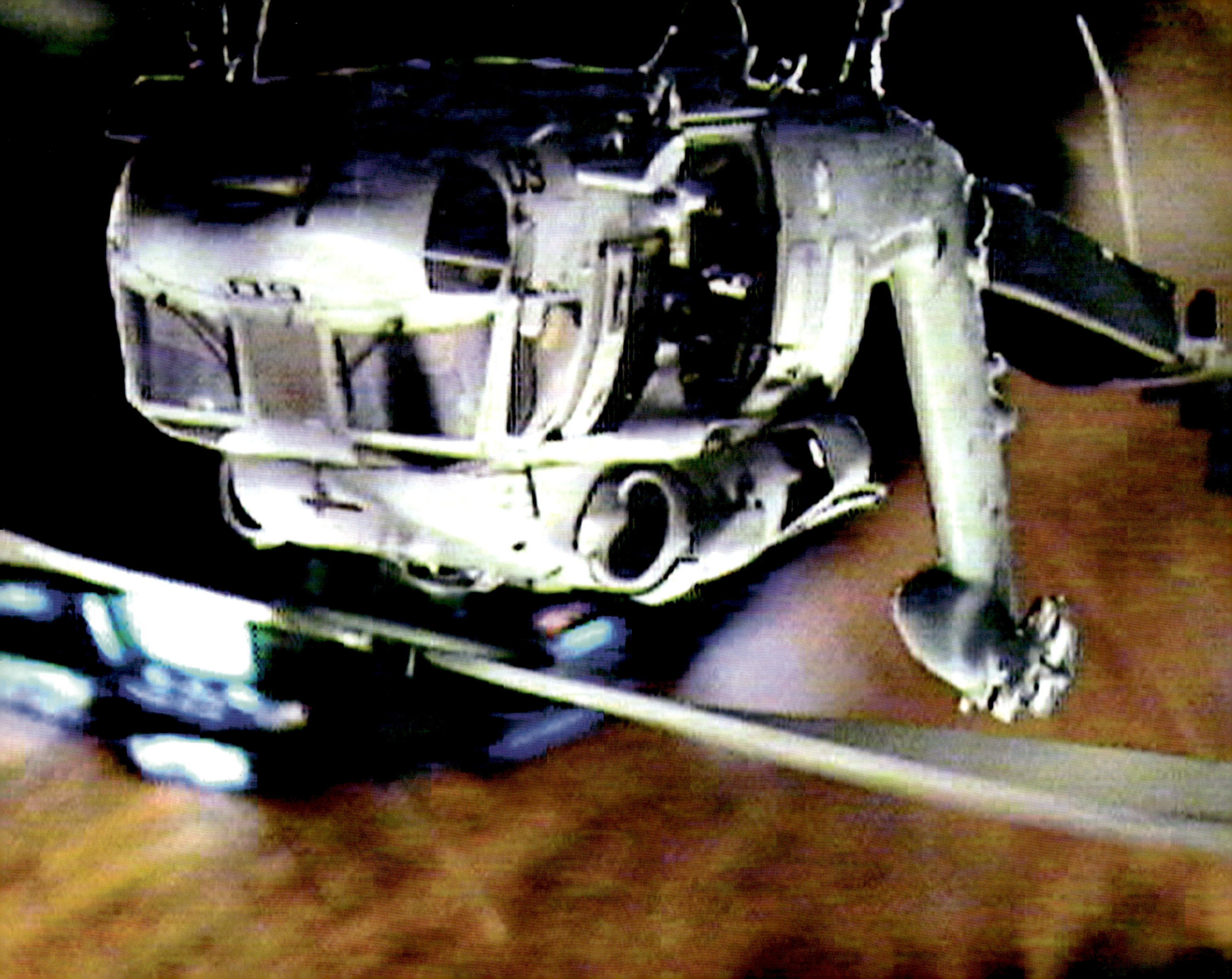

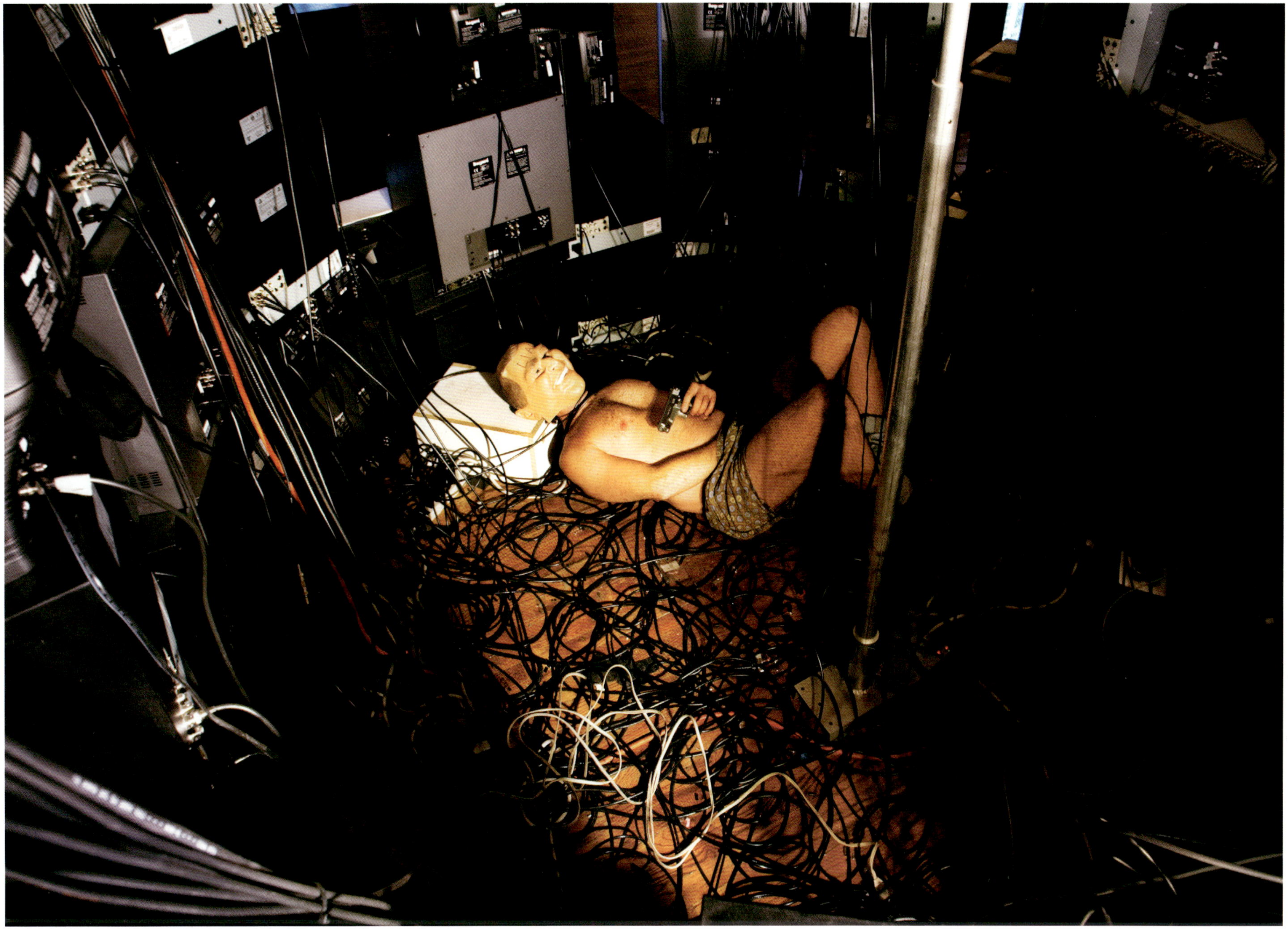

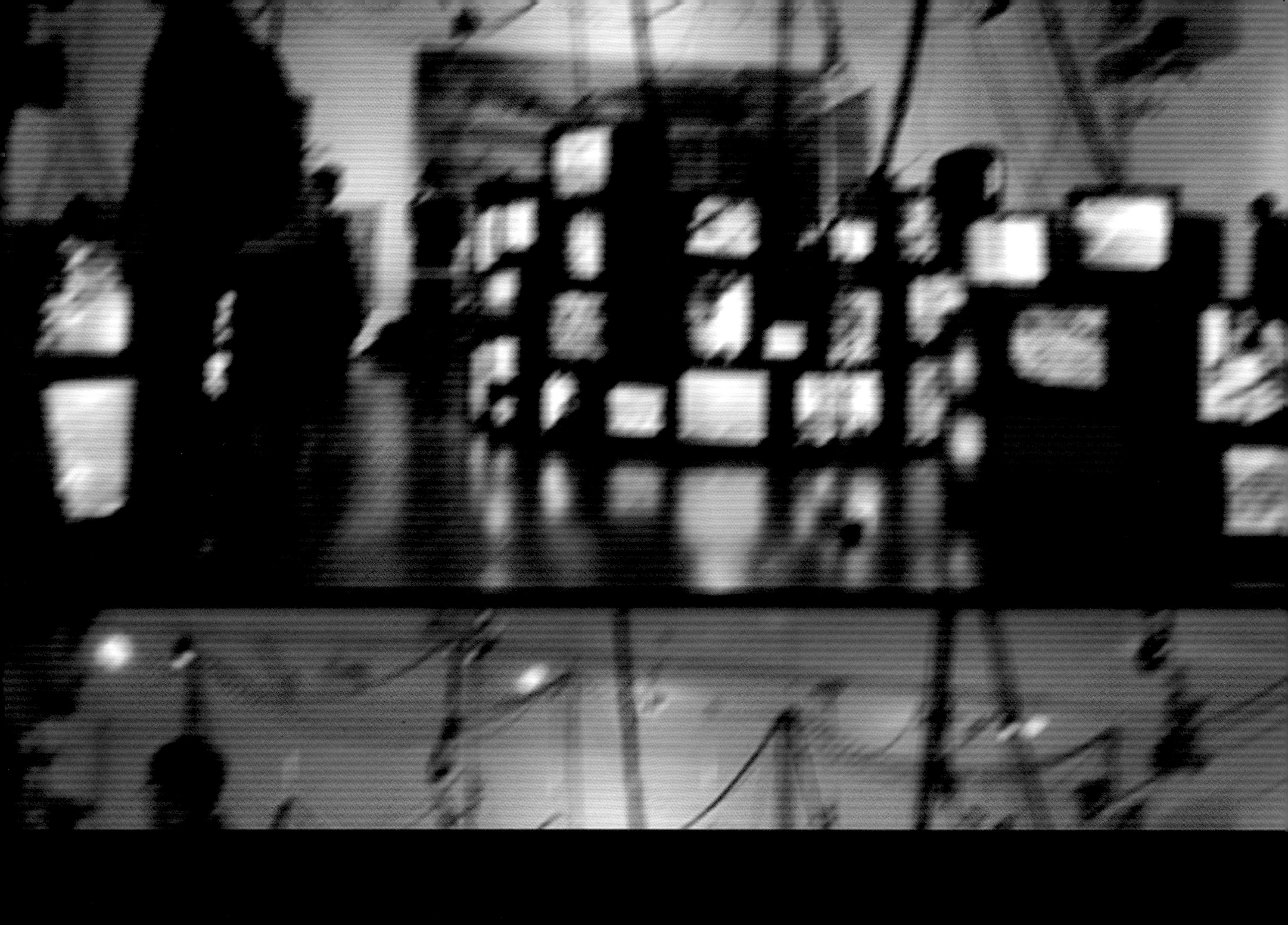

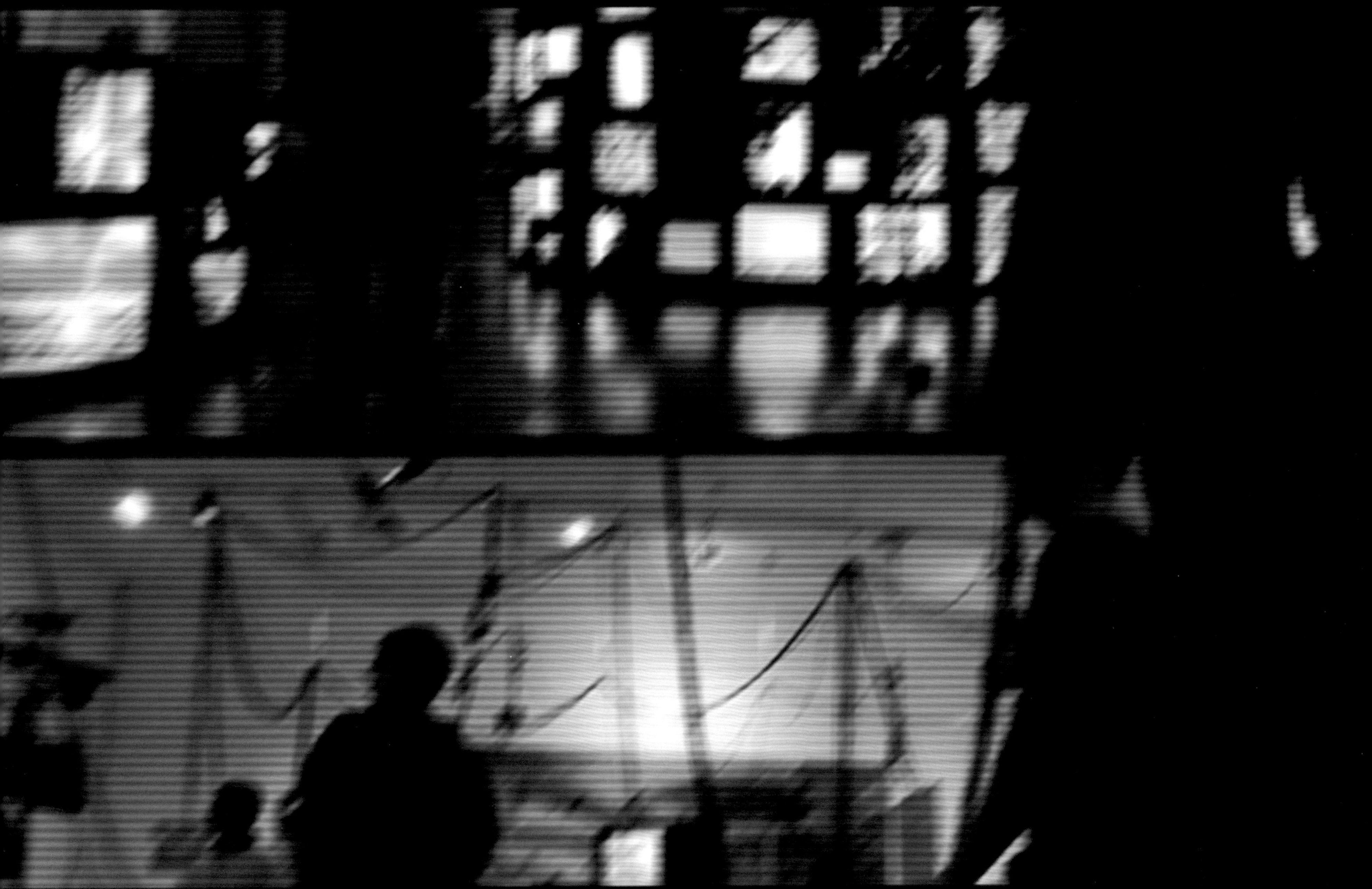

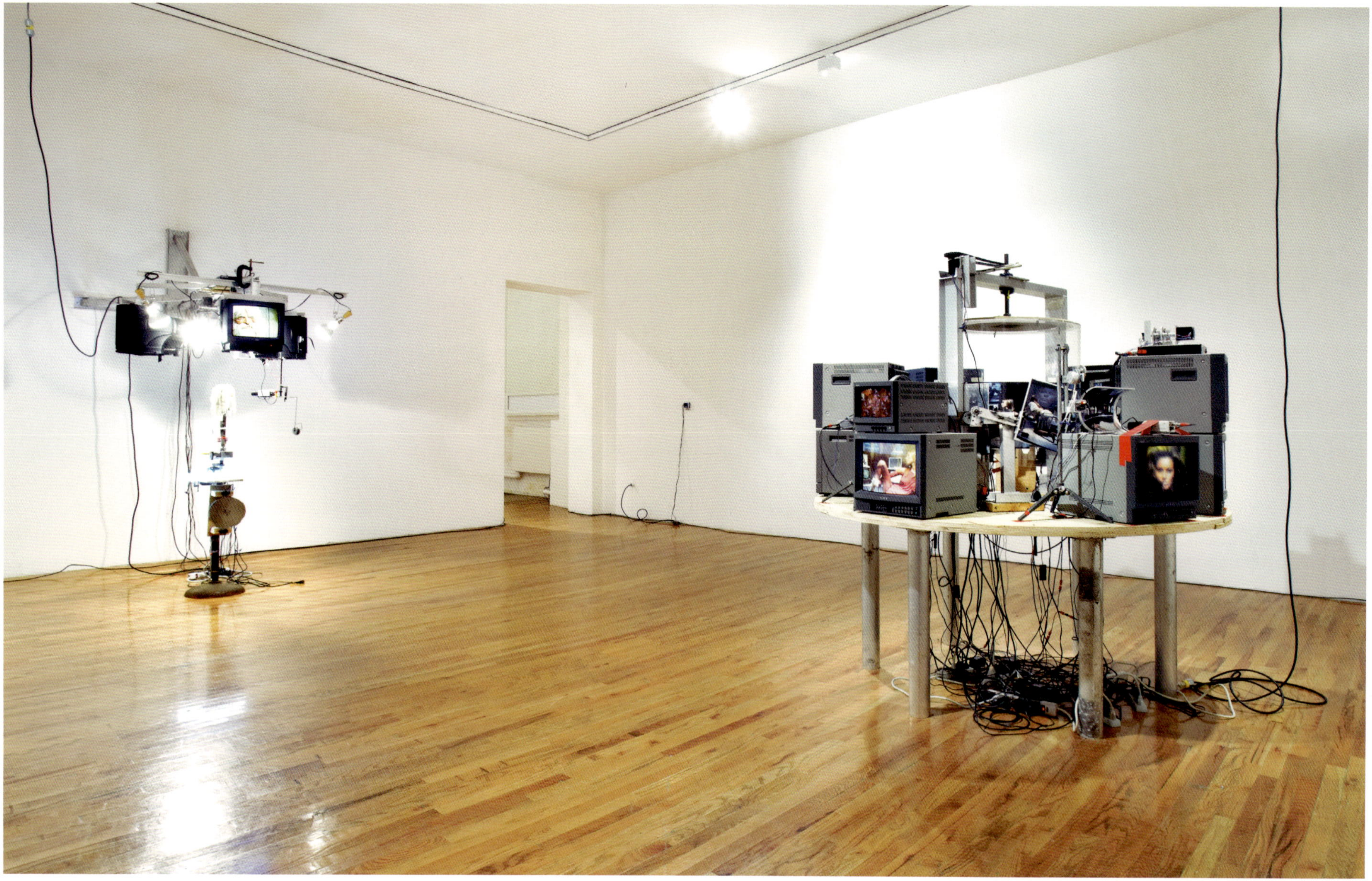

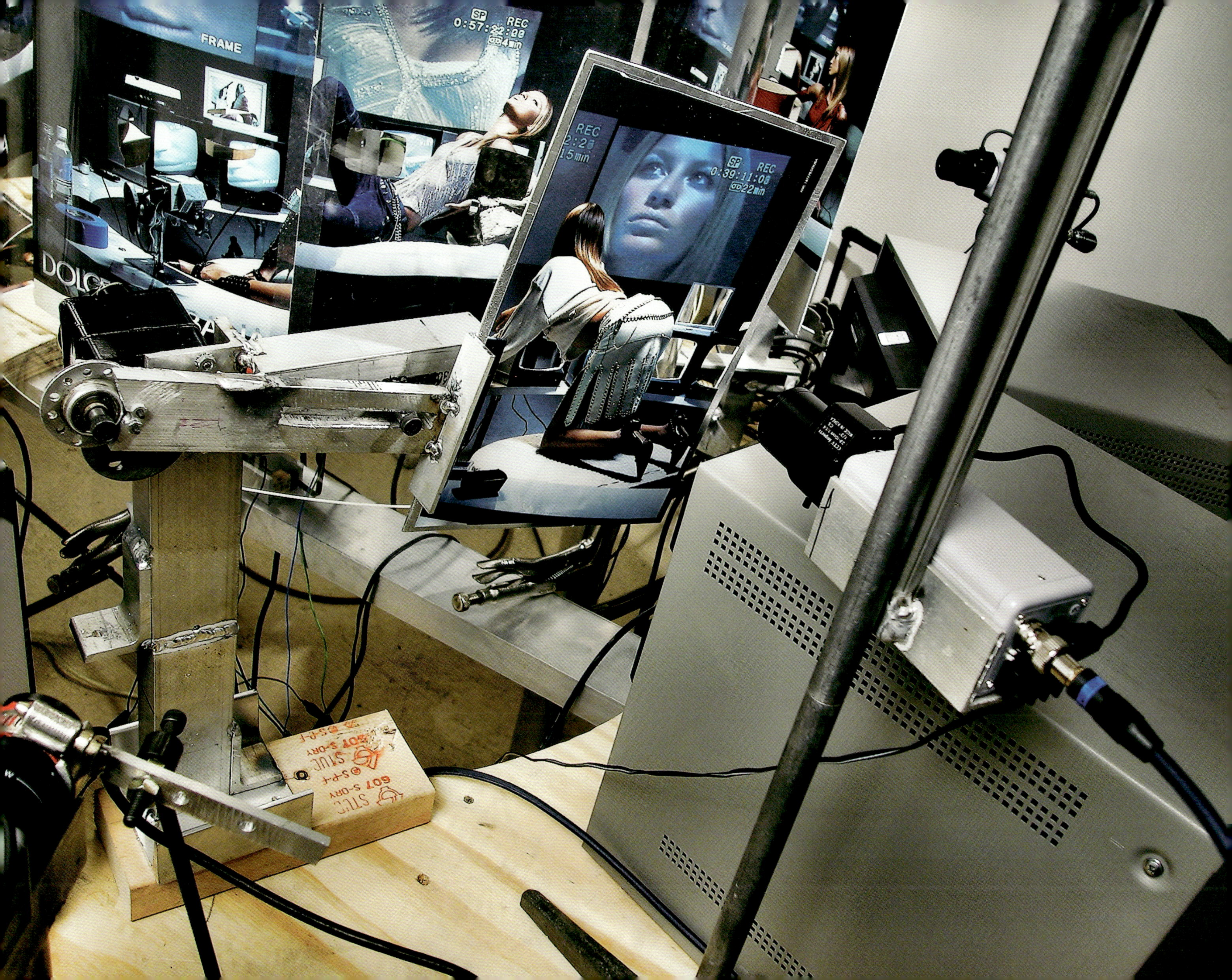
FRAME

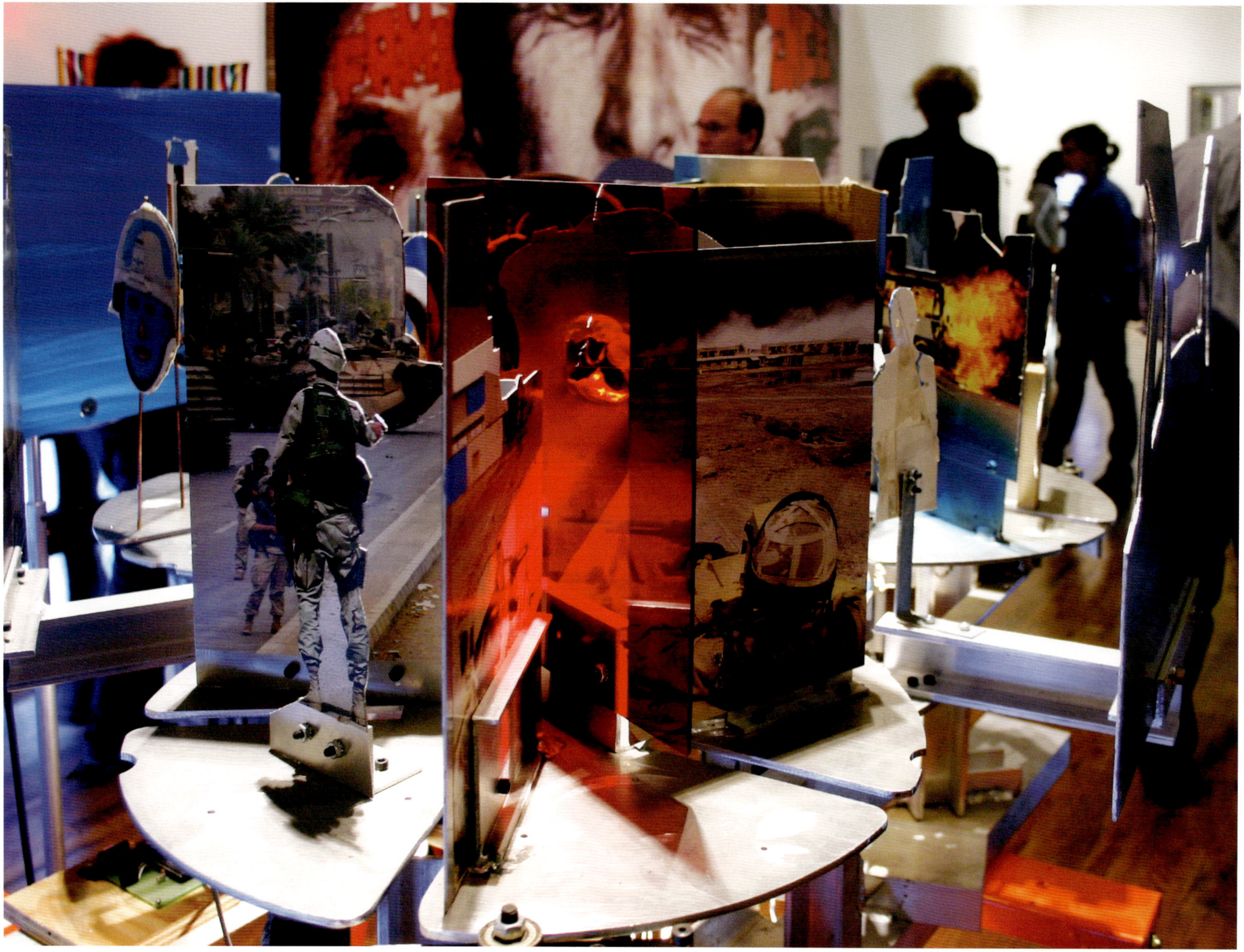

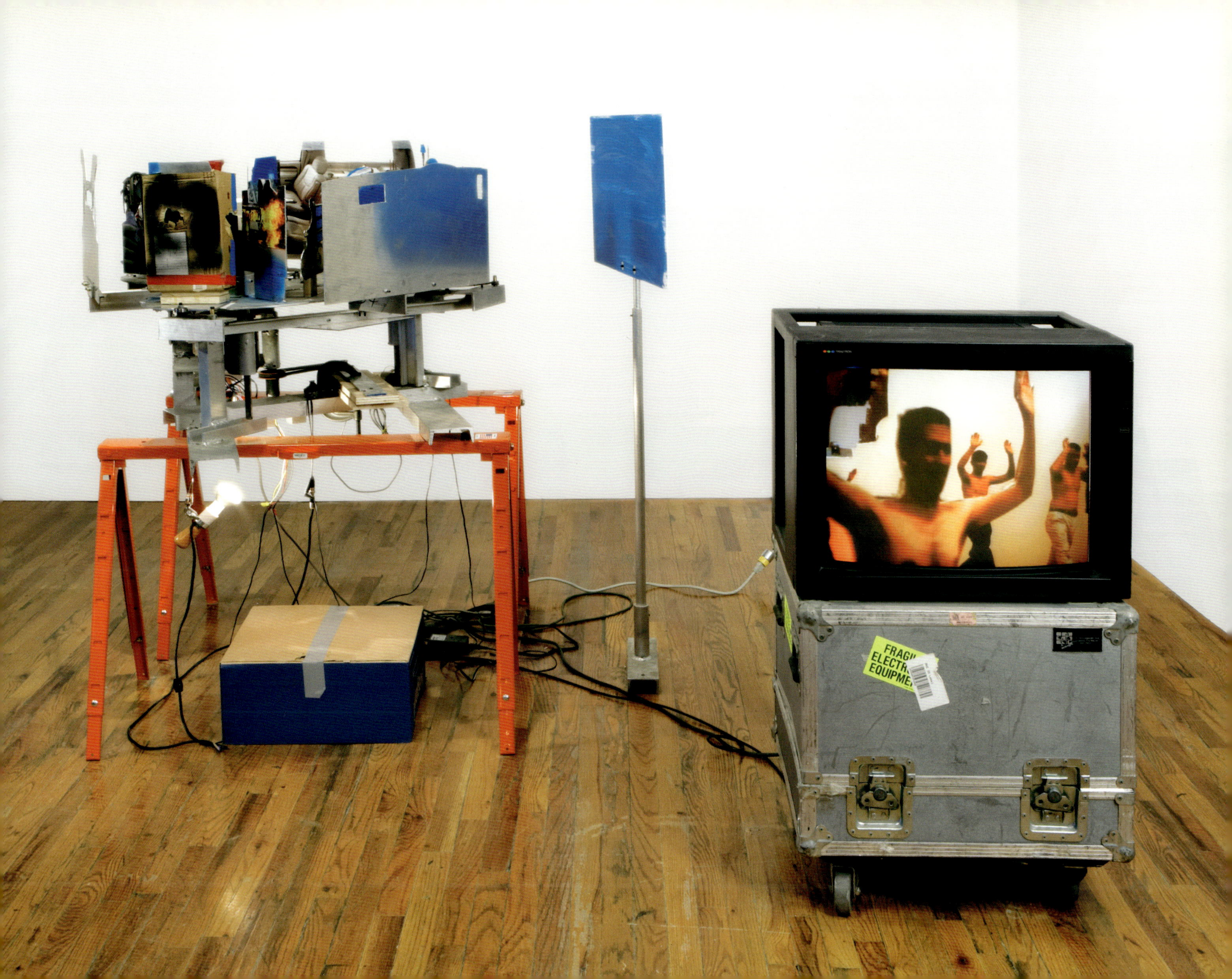

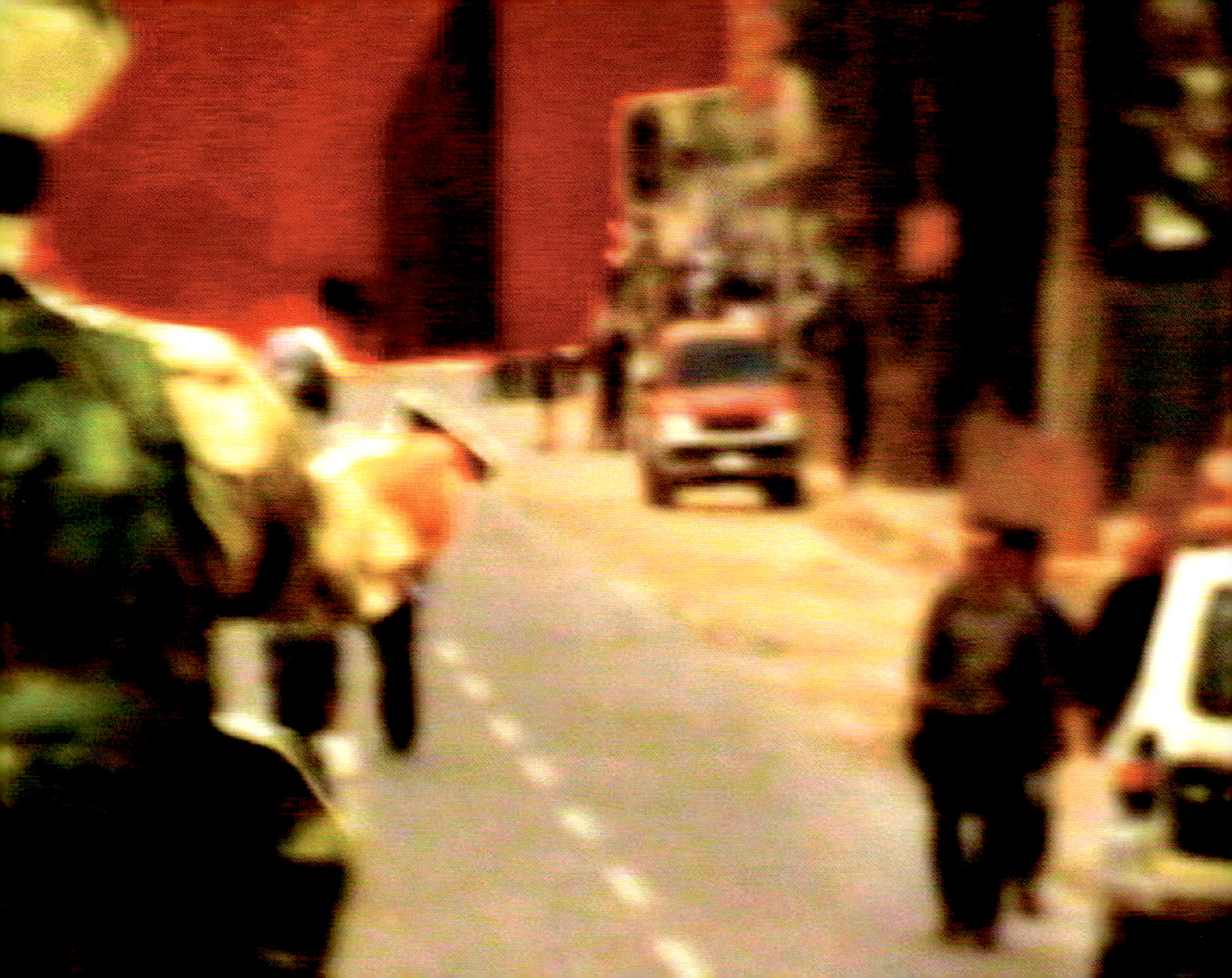

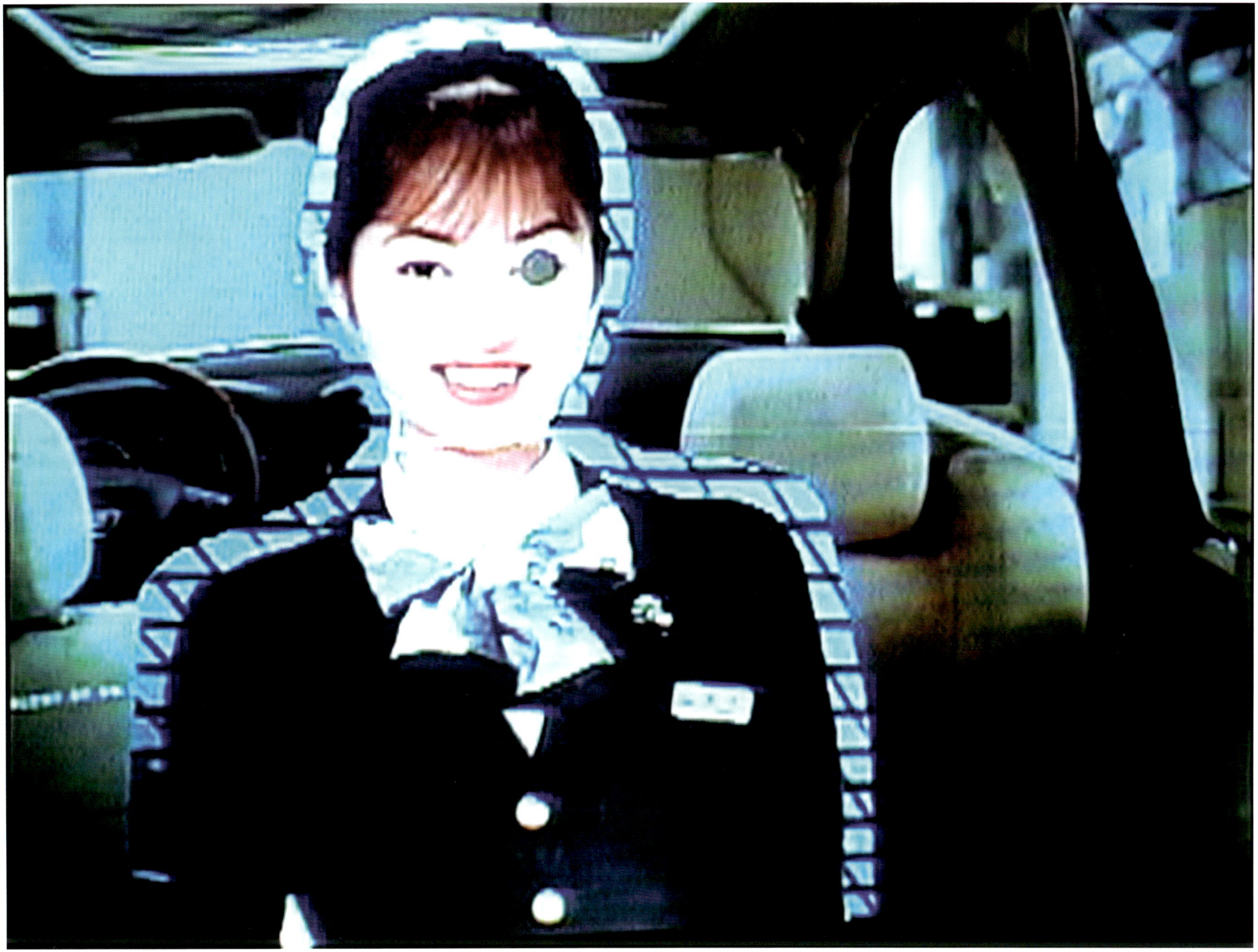

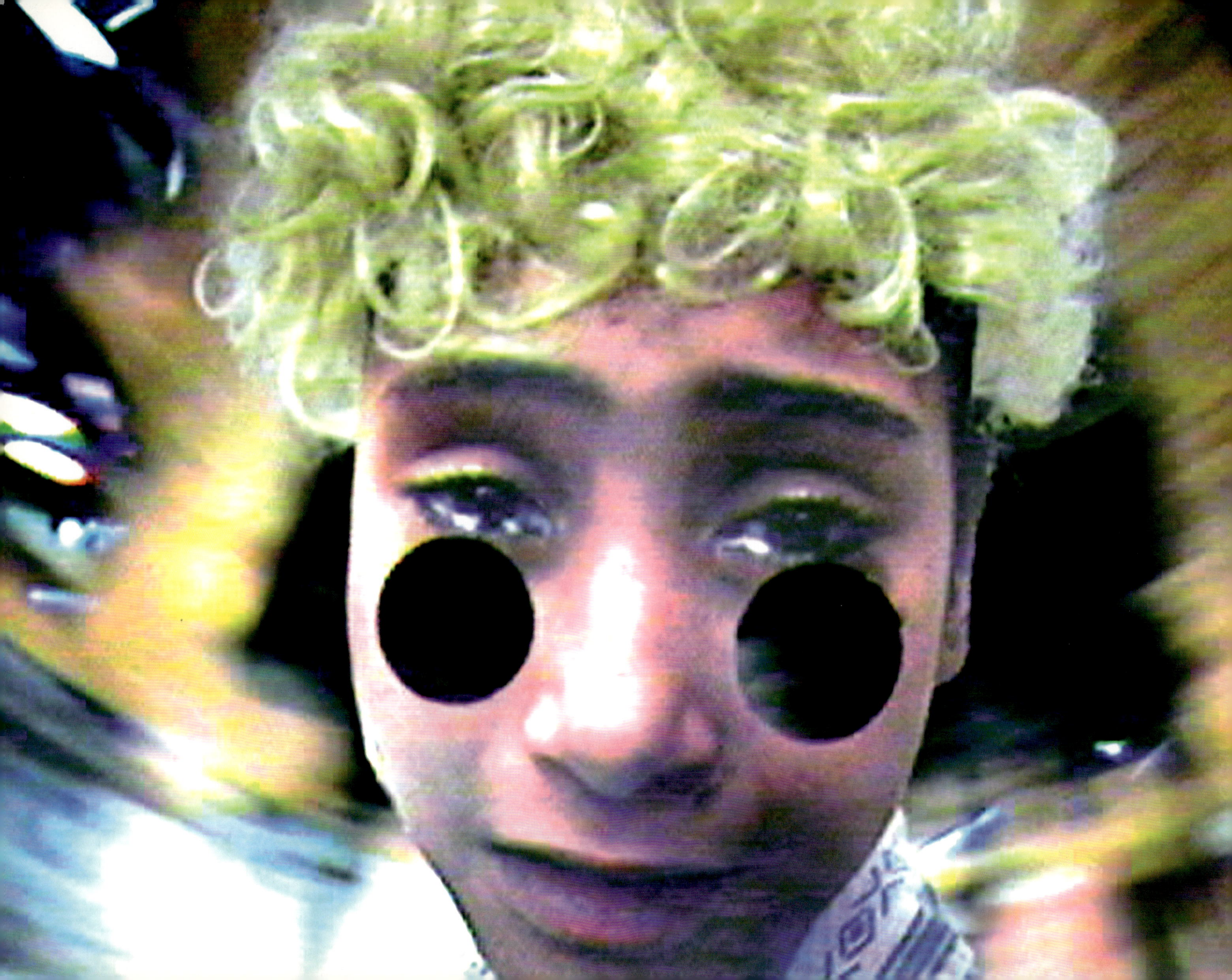

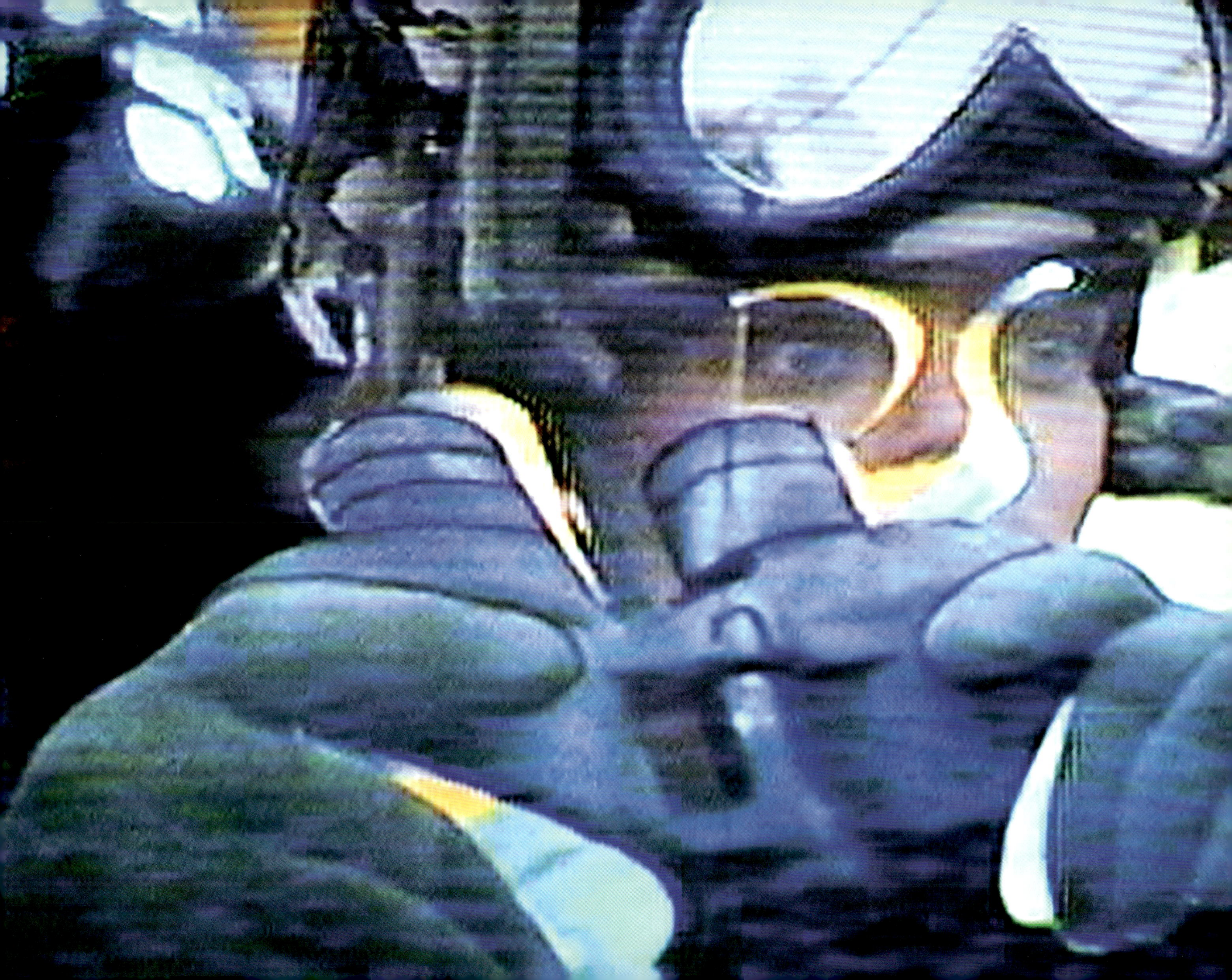

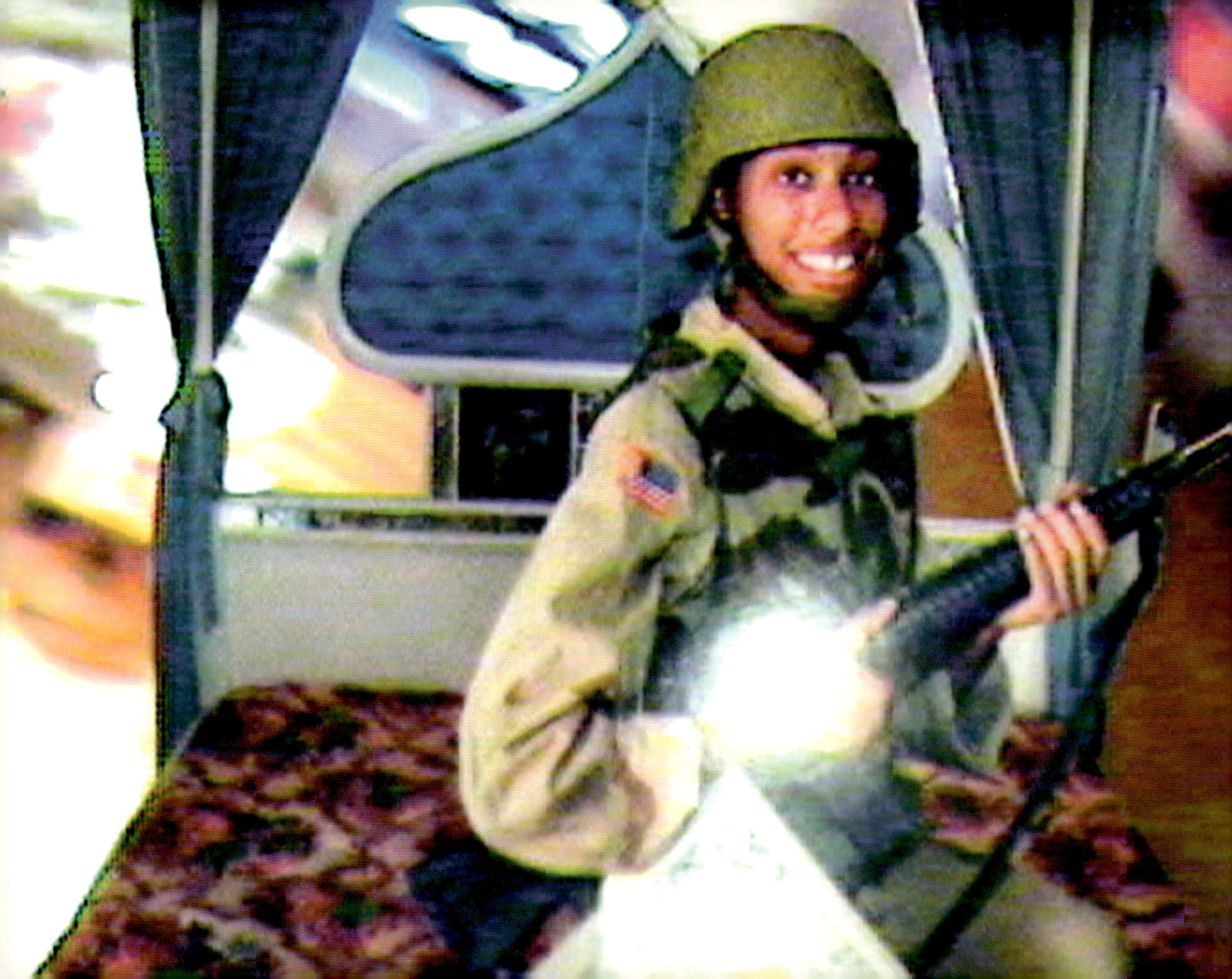

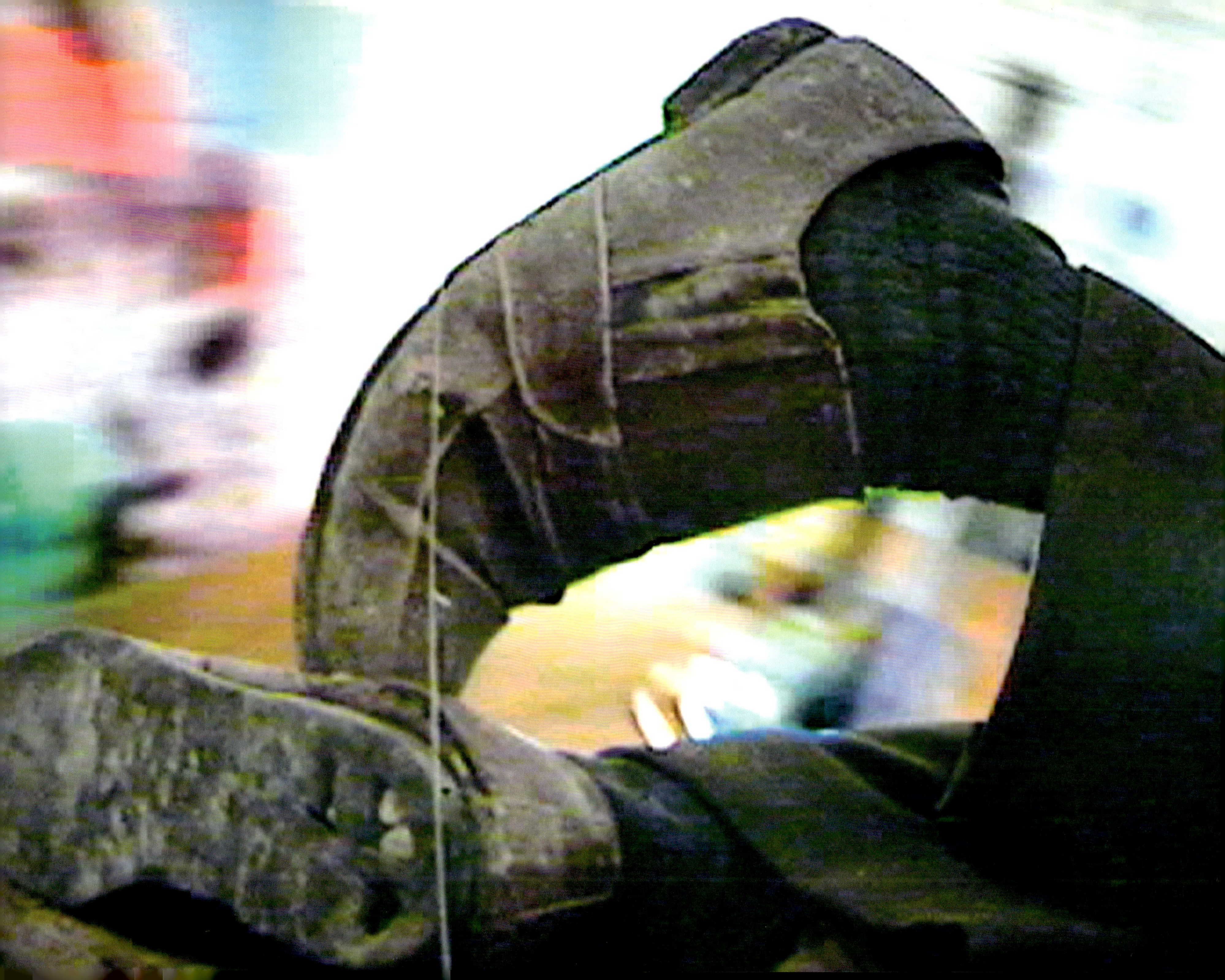

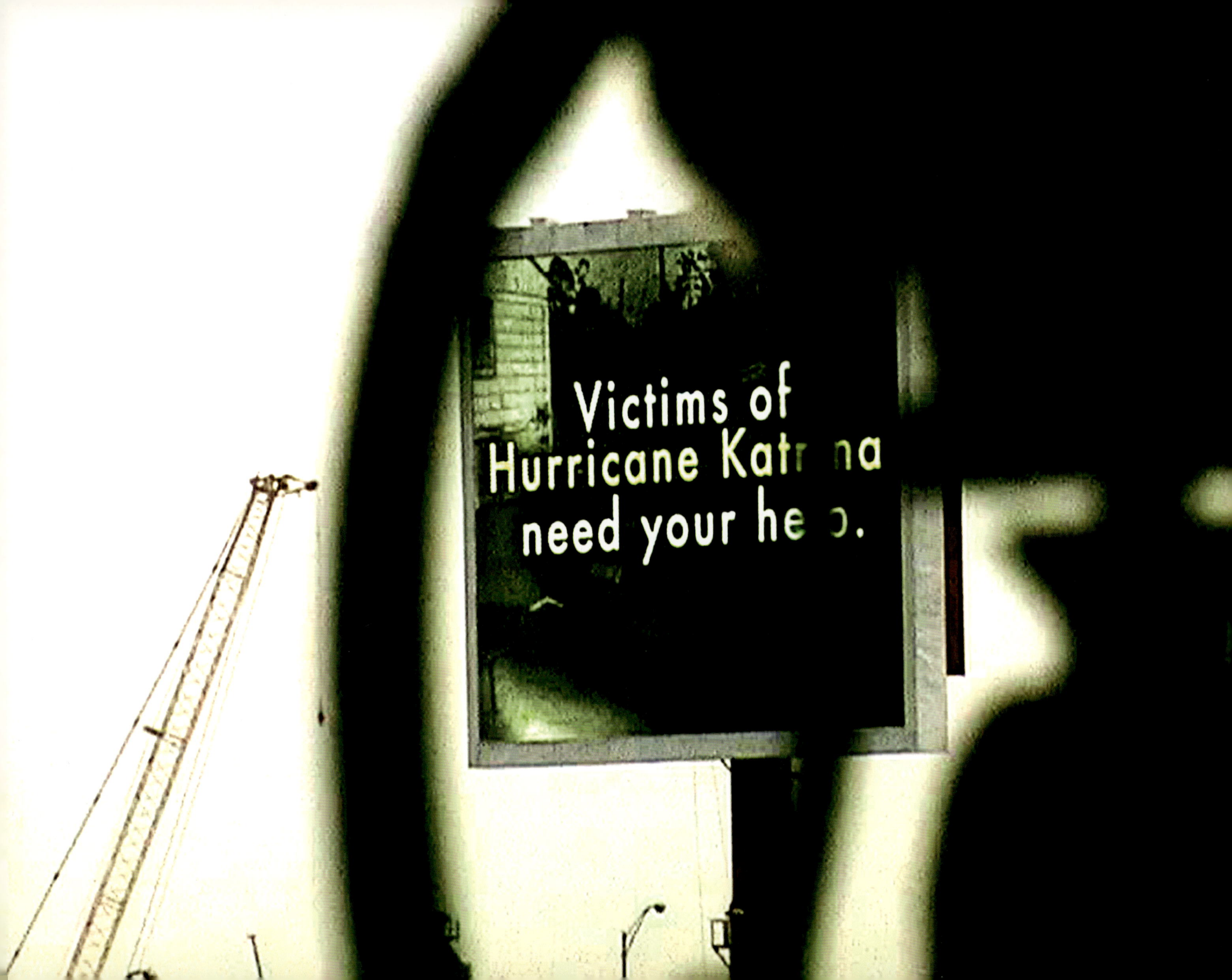
Victims of
Hurricane Katr na
need your he

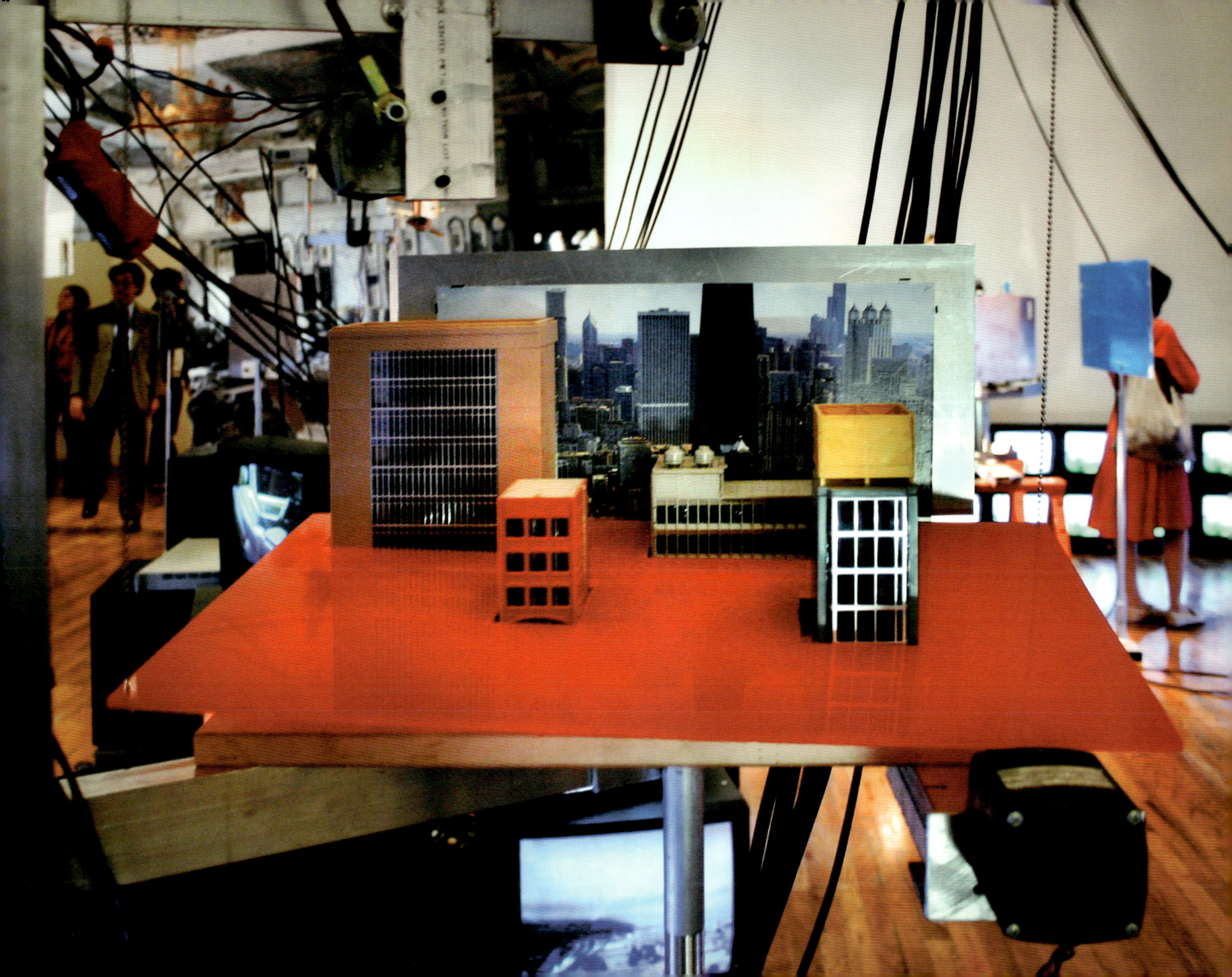

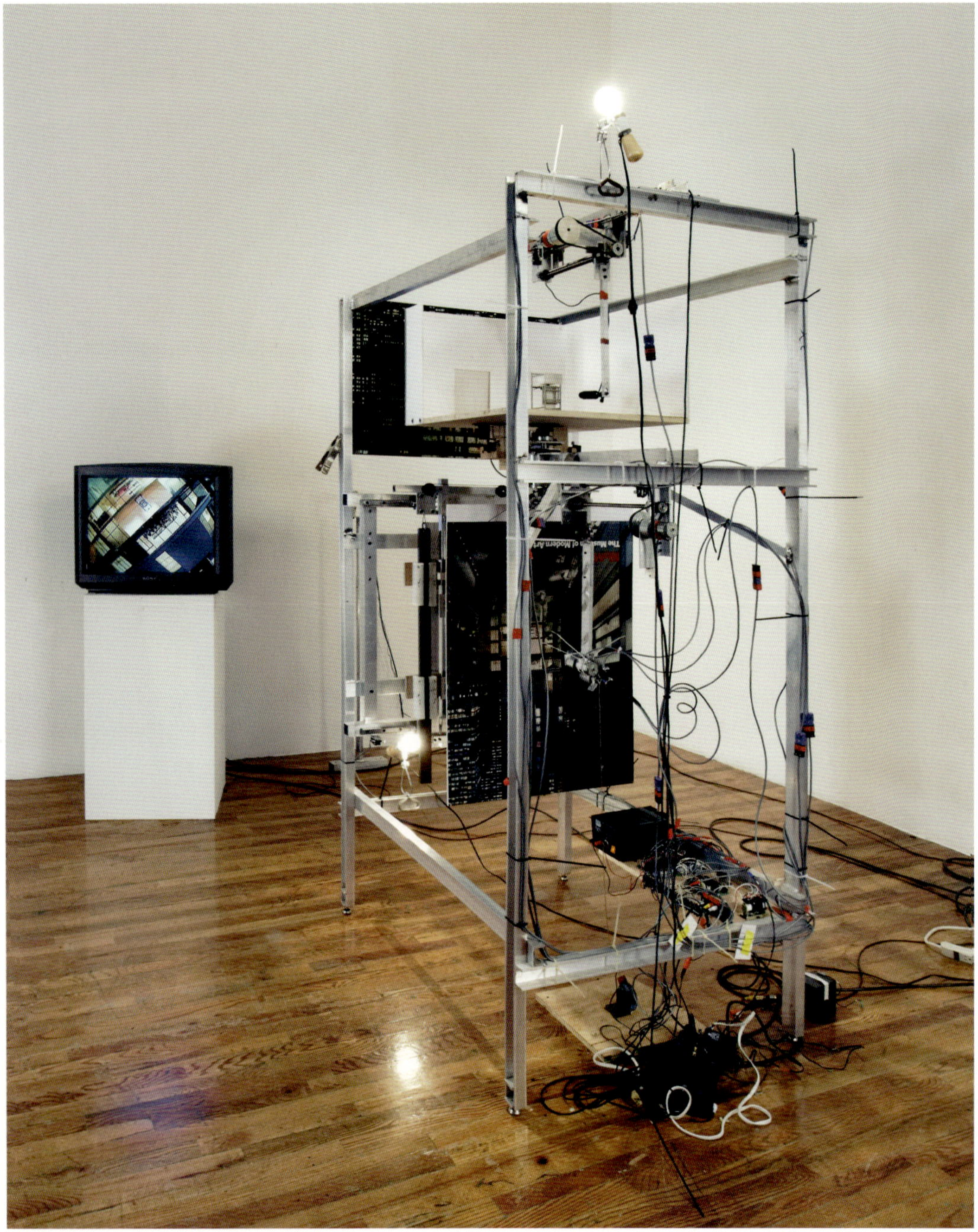

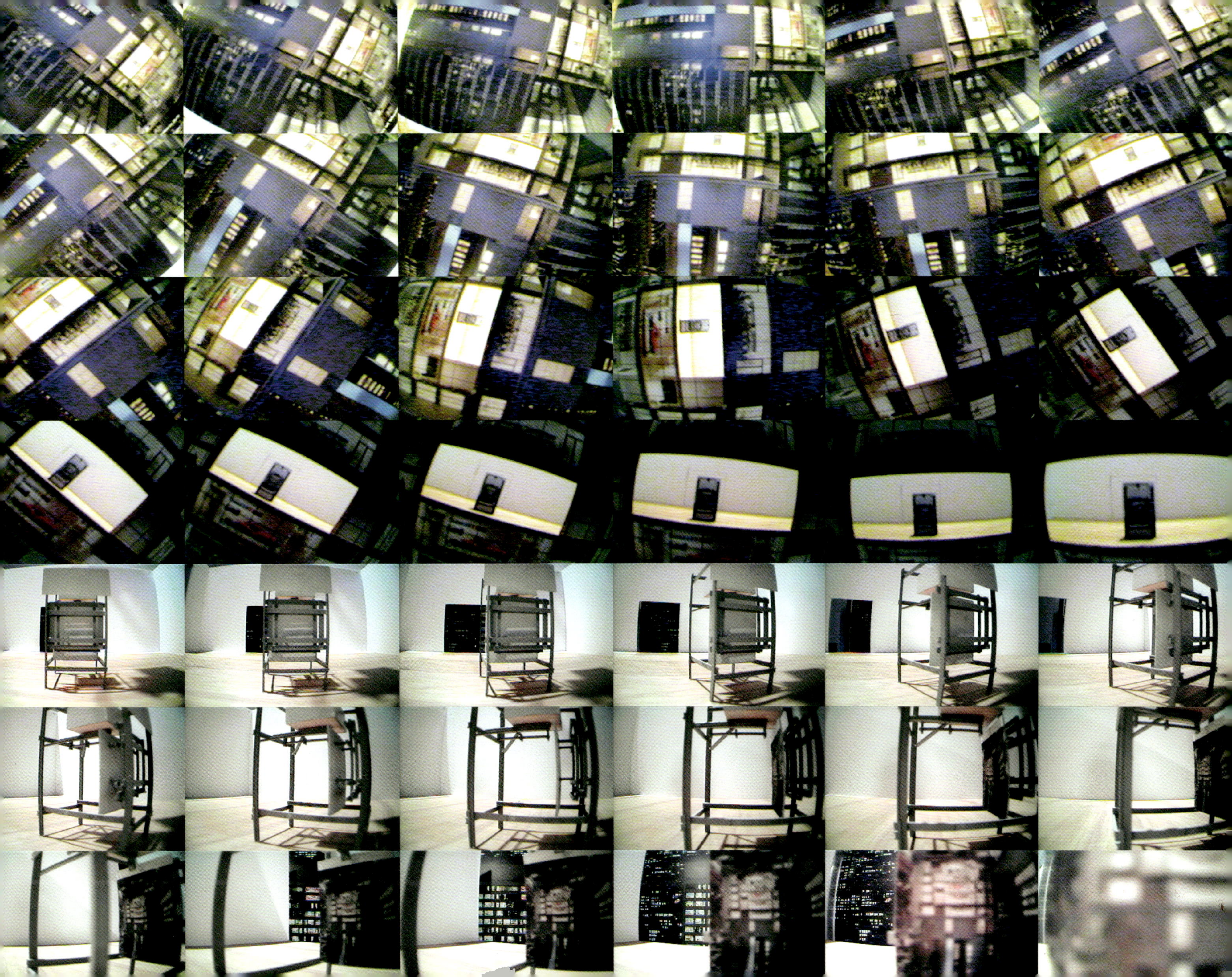

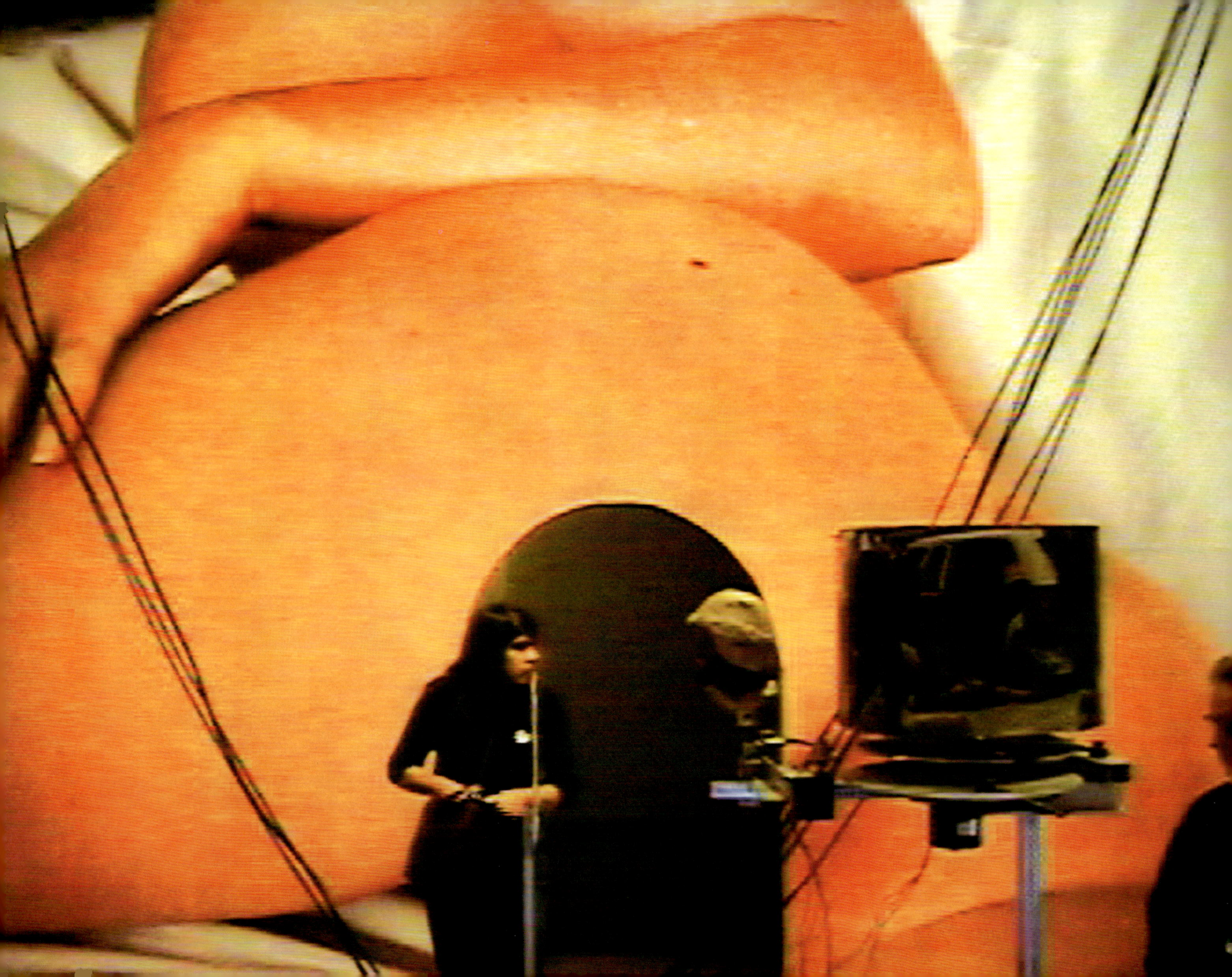

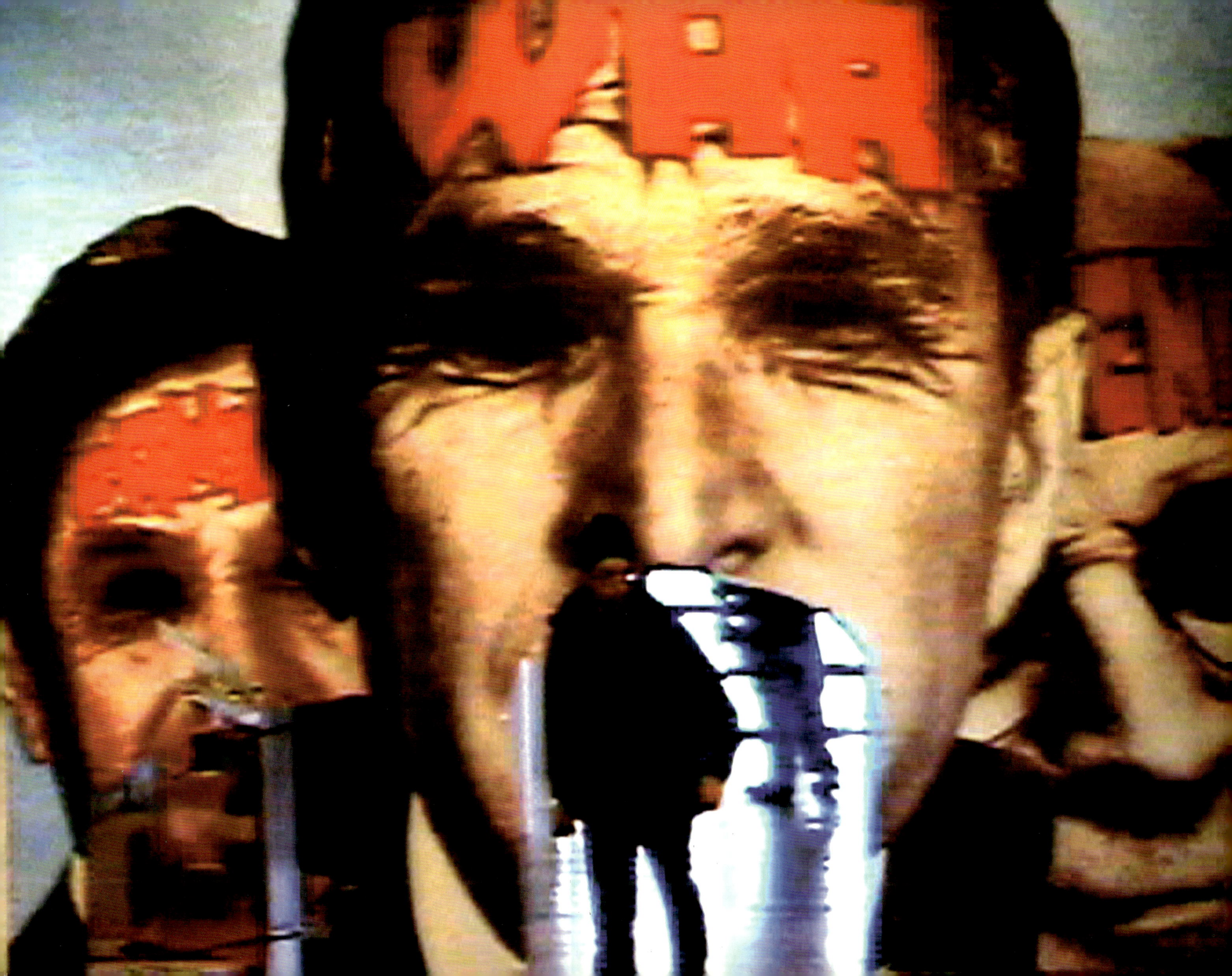

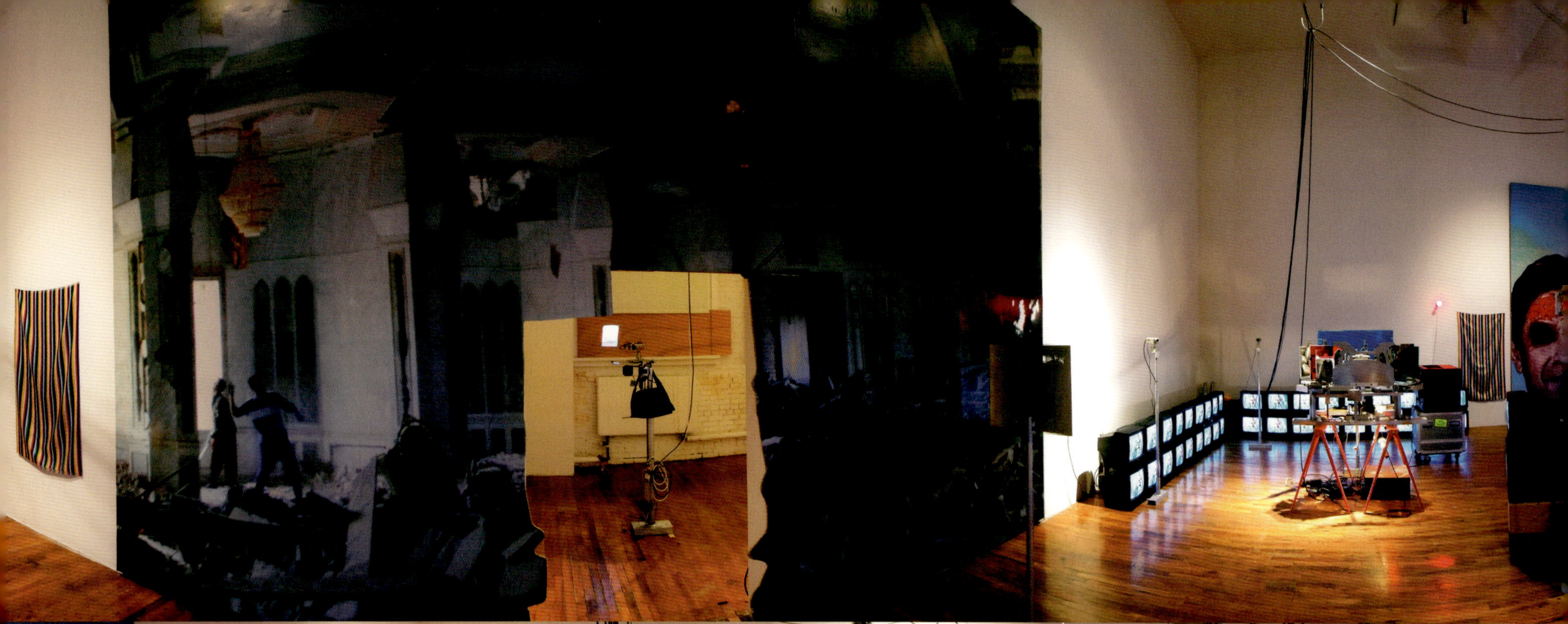

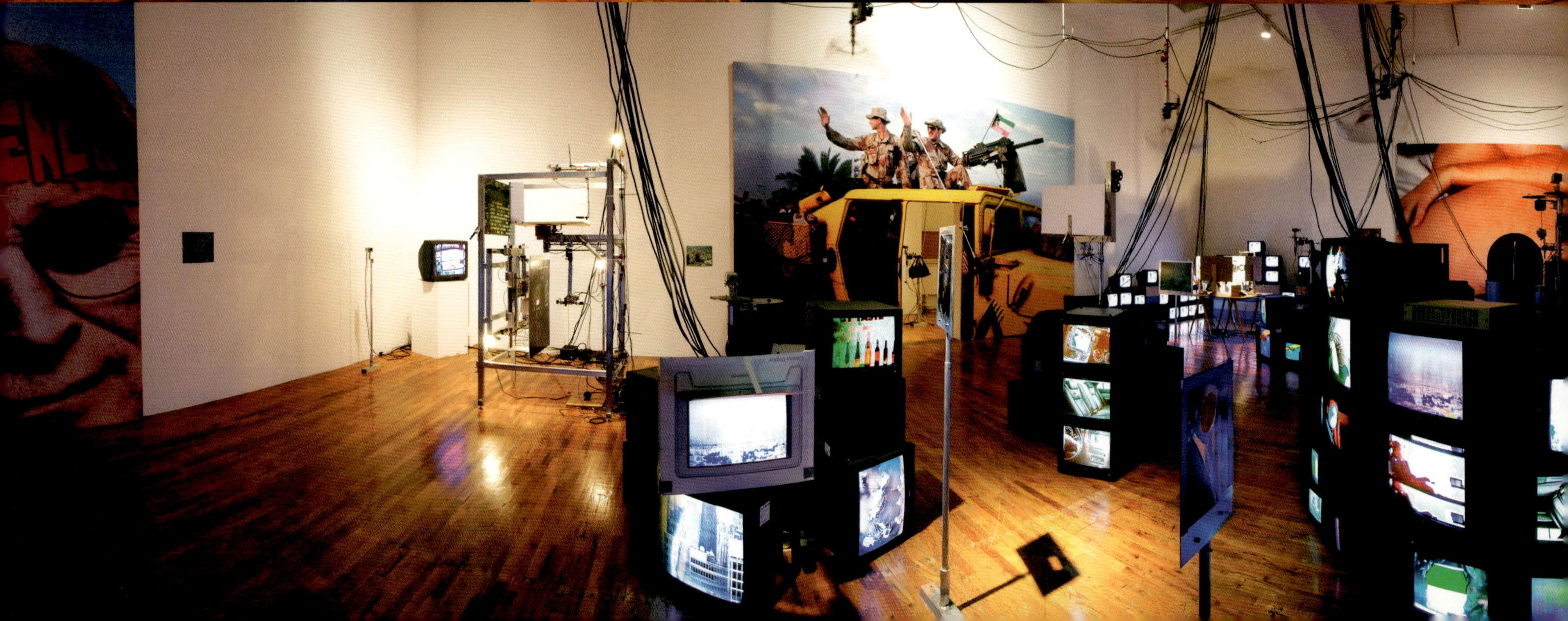

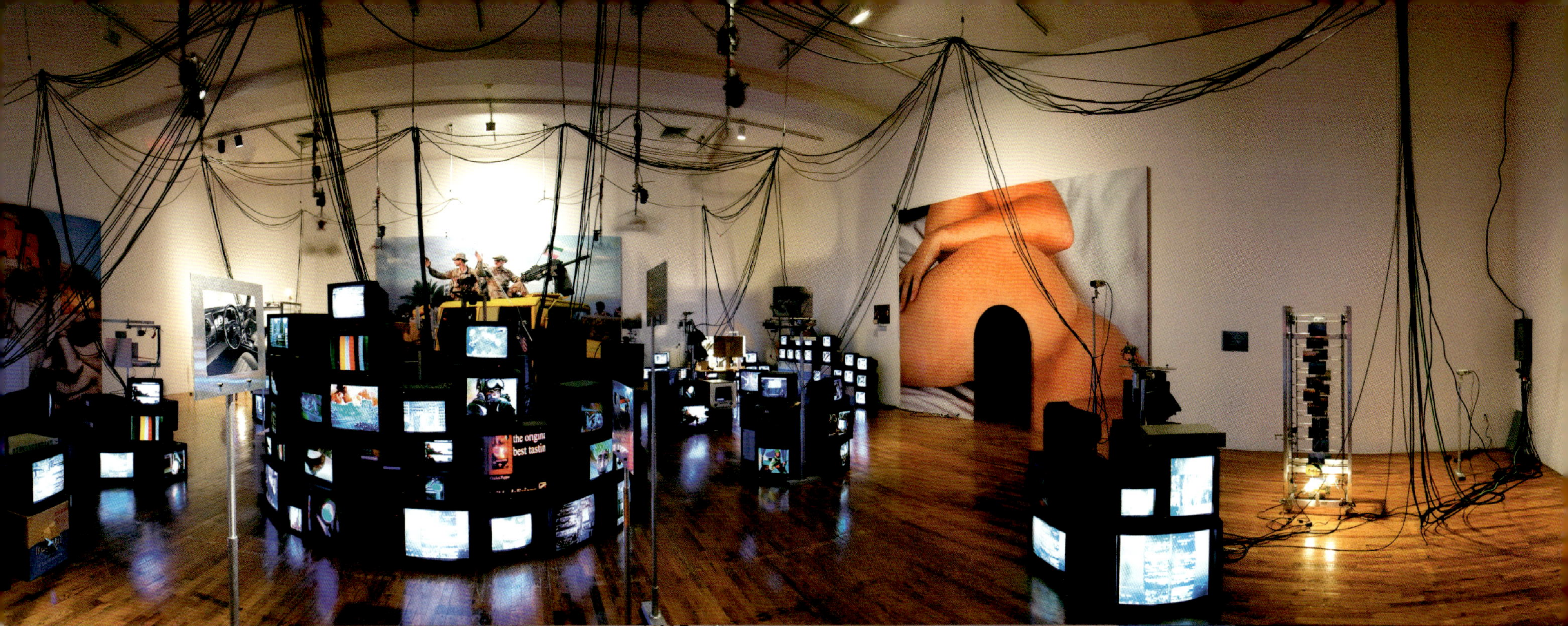

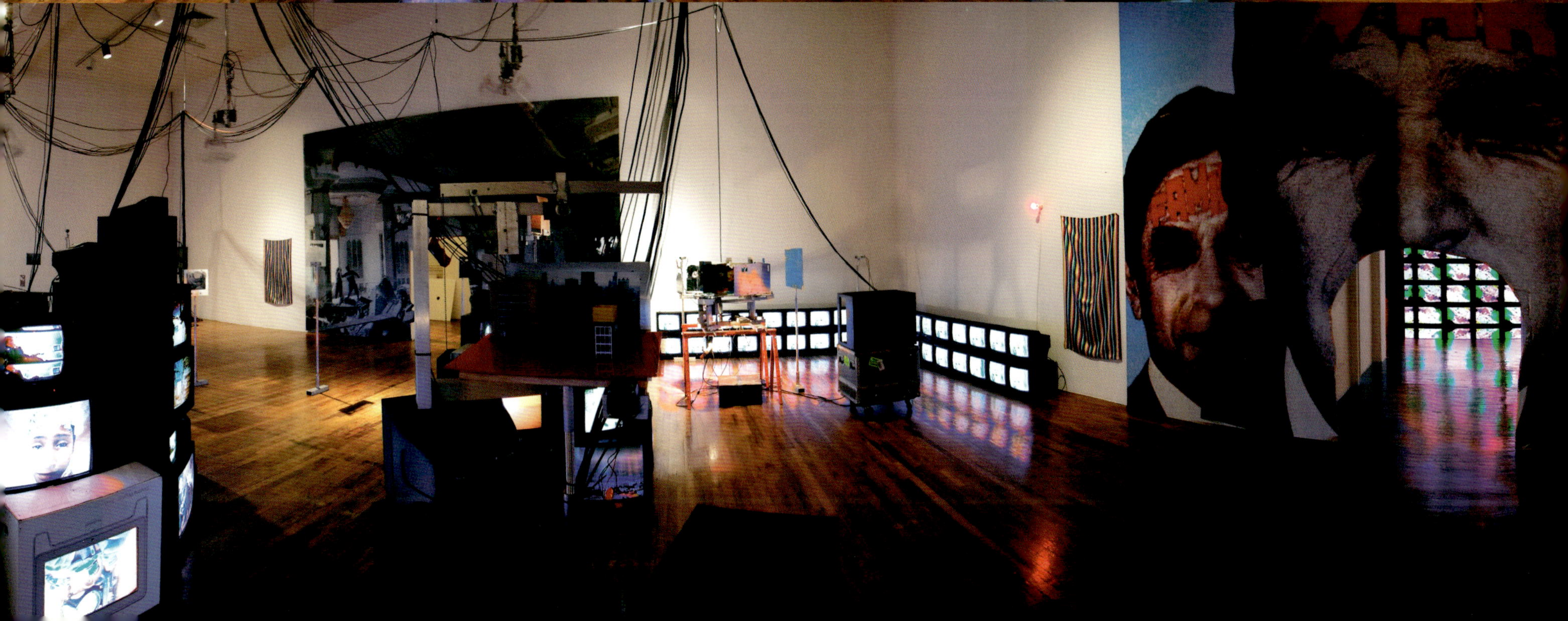

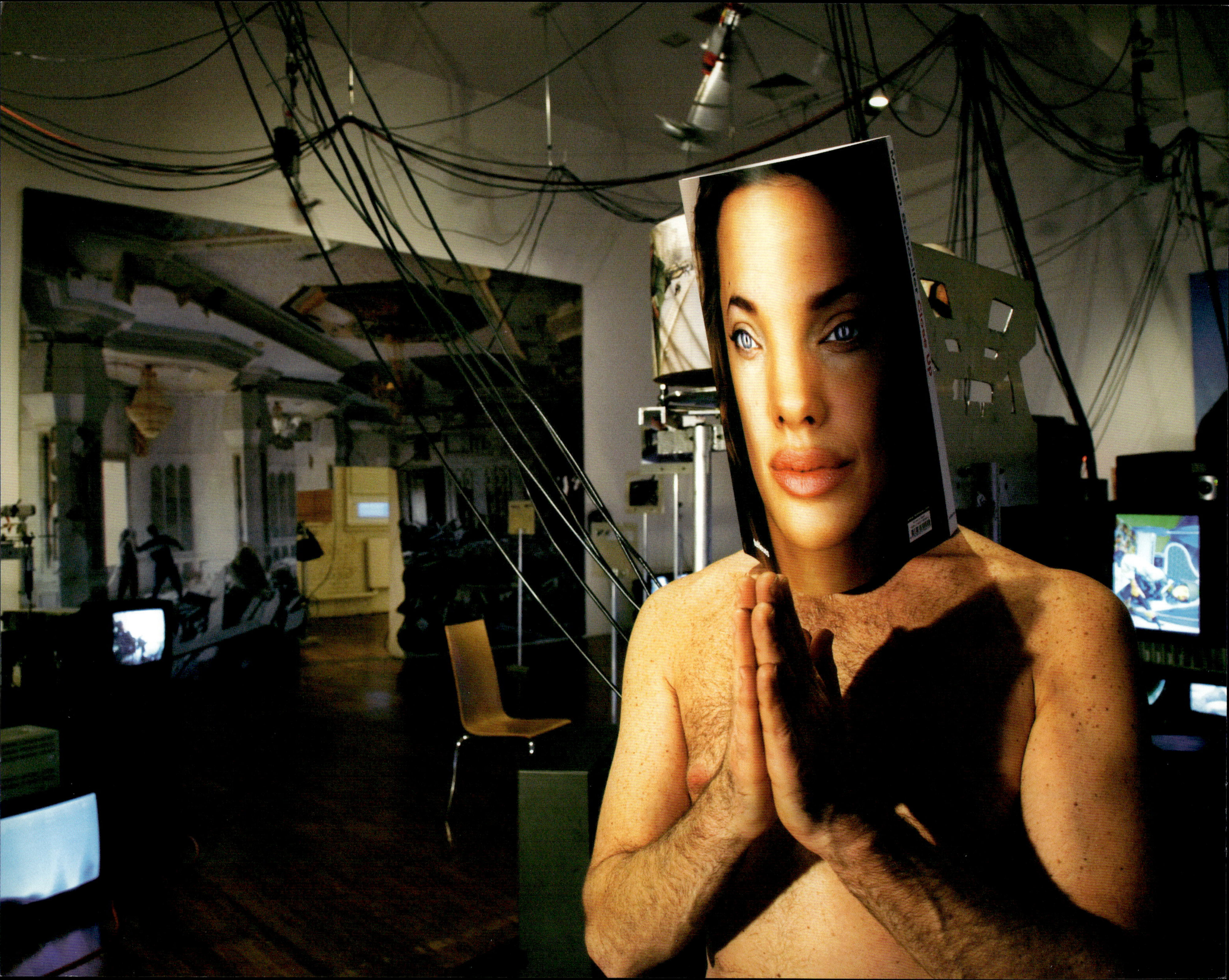

Alanna Heiss

The Palace at 4 A.M.

Part of the Baby-Boomer, TV generation who would go on to address the "image world" they were born into, Jon Kessler is best known for kinetic sculptural works that deconstruct and reconstruct a myriad of culturally recognizable objects and imagery. Kessler first came to prominence in the mid-1980s, exhibiting widely in the United States, Europe, and Asia. He participated in many of the important exhibitions at the time, including The Museum of Modern Art's "International Survey of Recent Painting and Sculpture" in 1984; the 1985 Whitney Biennial; and "Endgame" at the ICA in Boston, in 1986. Also in 1984, one of the artist's early works was shown in a Project Room here at P.S.1.

In 2004, Kessler presented "Global Village Idiot," a suite of mechanical sculptures, at Deitch Projects in New York. This suite of eight works mirrored a post-9/11 sensibility, fashioning a dystopic reality in which life was inextricably linked to a sense of anxiety, fragility, and terror.

"The Palace at 4 A.M." is the largest and most complex of Kessler's endeavors and comprises a vast expansion on "Global Village Idiot." Referencing the 1932 Alberto Giacometti sculpture of the same name, the installation's title means to evoke, in the artist's words, "the insanity that happens at that hour when no one is watching." Sources of inspiration and information vary tremendously—including Kabuki, Francis Ford Coppola's *Apocalypse Now*, and reality television shows such as *The Swan*—but Kessler's work ultimately speaks to the history of representations of warfare as well as to that of global image production.

The P.S.1 installation encompasses a network of kinetic sculptures, each of which incorporate surveillance cameras acting in tandem with their movements to create video imagery occurring in real time. In the main gallery is a veritable barrage of visual stimuli. Actions and images created live in peripheral galleries, as well as in the main space, are simultaneously displayed on monitors and

Alanna Heiss
Director
P.S.1 Contemporary Art Center

screens in this central "nerve." Transmitted information is constantly assembled and reassembled, resulting in scenes that appear unfamiliar, disorienting, and sometimes threatening.

One of the tenets of Kessler's practice is his uncompromising "exposure" of the apparatuses used to produce his imagery. The wires, gears, cameras, and motors that "fuel" his devices are always visible and in large part constitute the bodies of the works themselves, exploding accepted notions of inner and outer, of what is seen and what is unseen. Though the physical structure of sprawling installation is seemingly demystified, we nevertheless remain compelled to question the enigma of the image.

The "Palace at 4 A.M." is a large, multi-part sculpture that traces and records this time in our society from the vantage point of a great visual artist. I am proud to have been a part of its exposure to the members of the society it so beautifully and tragically documents.

Hal Foster Pandemonium

No artist has produced a more pointed, or more powerful, response to the events of 9/11 and the debacle of the "War on Terror" than Jon Kessler. With image-machines that are at once deeply funny, deadly serious, and wholly anarchistic, Kessler takes bites out of the spectacle of the Bush regime—signal images of its authoritarian politics, misbegotten wars, and clueless culture—chews them over, and spits them out again. The effect is one of calculated rage against the current machine of American Empire.

Kessler is well known for his homemade mechanisms that activate found representations, usually drawn from mass culture, often with delirious lighting and compulsive movement. Yet over the last five years—that is, since 9/11—a shift has occurred in his work. He has introduced video, mostly in the low-tech form of small surveillance cameras, some of which relay the bizarre actions of automatons on nearby monitors. He has expanded the scale of his mechanical tableaux, sometimes to the point where they almost engulf the viewer in a noisy tangle of gadgets, screens, cables, and wires. And he has responded, directly and indirectly, to the image-world of the Bush era, reworking news bites, military reports, tourist postcards, seductive ads, and franchised toys into delirious little dramas that deconstruct some of the political fixations and cultural fascinations of contemporary America. Imagine *The Light-Space Modulator* of Moholy-Nagy redone with gizmos found on Canal Street by an artist who (like Malcolm McDowell in *A Clockwork Orange*) was forced to watch the awful events of the last five years on television, and that might conjure up some of the effects of his two recent installations, "Global Village Idiot" (2004) and "The Palace at 4 A.M." (2005–06).

Even when the image-machines of "Global Village Idiot" were related thematically, they remained discrete physically. This is less the case with "The Palace at 4 A.M.," which absorbs some stations of "Global Village Idiot" and develops others into a vast, disjunctive environment. Two of the most provocative stations in these installations are *One Hour Photo* and *Heaven's Gate*. In *One Hour Photo*, tacky postcards of the World Trade Center revolve on a vertical conveyor belt, at the bottom of which is placed a small video camera. As each card approaches the camera and then trips over it, an image is produced on a monitor of a wobbly zoom toward the Twin Towers. This perverse point of view, in which we are led to identify with what is effectively a weapon, first became familiar during Gulf War I, with the little cameras located in our "smart bombs." In *One Hour Photo*, Kessler conjures up the perspective of both terrorists and victims on the doomed planes—a point of view obscured by the repeated videos of the jets seen from the ground. In effect, Kessler, who lives only a few blocks from Ground Zero, offers up an imagining of this other trauma, an imagining that, as a Freudian might say, cannot be worked through psychologically and so must be repeated compulsively. In this way, he also captures an essential aspect of spectacular events of this sort—how they are at once traumatically real and utterly mediated.

Just as provocative is *Heaven's Gate*. The video in this station flies through a model city into a model apartment, where it zooms in on a computer screen; here we see the ass of a doll, then pass through it, only to emerge, on the other side, into the space of a gallery—an anal rebirth into the art world. As David Joselit has commented of this piece, Kessler "imagines representation as a carnal act": In opposition to the pervasive myth of virtual information, we are asked to consider the material, often corporeal, effects of data (Joselit also notes that "Global Village Idiot" appeared in the midst of reports of the Abu Ghraib torture). More broadly, we are led to reflect on a general condition of obscenity, in contemporary news and entertainment alike, in which representations, bodies, and machines often converge violently.

Kessler works by blunt appropriation and perverse refunctioning, and this goes for his titles, too. "Global Village Idiot" is a clever contraction of "global village" and "village idiot." The second term requires no explanation, while the first is drawn from Marshall McLuhan, who, in such texts as *War and Peace in the Global Village* (1968), argued that the instantaneous reach of mass communications had, for the first time in history, projected a planetary audience. The Kessler phrase suggests that in the global village of today, one of endless infotainment controlled by a handful of governments and corporations, we are actively trained to be village idiots—cultural know-nothings and political incompetents. This insight—that technological progress and social regression can be complements, not opposites—is key to the critical charge of his art, and it flies out in many directions at once. For Kessler is the idiot here; so is the viewer; Bush is as well (one is reminded of the bumper sticker "Somewhere in Texas a Village Is Missing Its Idiot"); and bin Laden is not excused either (in an interview, Kessler hilariously imagines Osama, holed up in a cave somewhere, watching *Nip/Tuck*, a television drama about the trials of cosmetic surgery). In other words, "global village idiocy" is an equal-opportunity condition from which no one is immune.

The use of video, Kessler has remarked, "freed me to think of the machine as events and the image created as the spectacle." This formulation points to the circularity of his image-mechanisms, but there are also breaks within them. For even as his machines stage events for his cameras, the setups are rough, and the viewer not only watches the low-tech images but also sees their madcap production, which is sometimes so close to destruction that the two cannot be easily separated. The automatic aspect of the image-mechanisms is thus far from perfect or stable: Like little Frankenstein's monsters, they almost threaten to turn, if not on their maker, then on their viewer. And this viewer is also far from whole or secure: One not only sees but also is sometimes seen, and no two viewers witness precisely the same thing. Machine and image try "to complete each other, which is impossible," Kessler comments, and so "a puncture" is produced between the real and its representation—a puncture that allows us to see through these setups and, in principle, to see through others in the world. Through his own little dysfunctional spectacles, then, Kessler suggests that the great spectacle of American power is also in trouble, that its wizards cannot maintain its theater of illusions forever, that wondrous new technologies are always haunted by awful new disasters, and so on. And in this way, he points to another crucial contradiction of the American Empire today: Even as its power goes unchecked by its allies, let alone by its enemies, its image, especially in the Middle East, continues to take a beating.

In "The Palace at 4 A.M.," Kessler pushes these concerns to a new level. One enters this infernal world through a passage that is later revealed to be a giant beaver-shot. ("I do want viewers to be reborn when they enter my show," Kessler says, "but not in a clean state.") One then encounters a big blowup of the trashed residence of Saddam Hussein; the palace in question seems to be his. Yet one also sees a huge image of Bush with the word "war" scrawled in blood red, so the palace might be the White House as well. In short, the palace seems to be a psychological bunker that they share—and that, too often, they have made the rest of us share as well. At first glance, this *Gesamtkunstwerk* of media overload seems to be a carnivalesque world, turned upside down and inside out, but it soon becomes clear that this is actually how much of our world is. Finally, then, the palace in question might be your own home, too, say in the middle of the night when the "War on Terror" troubles your sleep. Indeed, one experiences an oscillation between engulfment and recoil in "The Palace at 4 A.M." that is a little like a nightmare.

With these installations, Kessler recalls various predecessors: Robert Rauschenberg and his rambunctious combinations of media appropriations, Claes Oldenburg and his regressive theater of homemade objects, Jean Tinguely and his auto-destructive contraptions, and so on. Closer to the present, one might also think of Mike Kelley and his inspired reenactments of the weird things that asocial men concoct in their basements and backyards. Other associations come to mind as well—media theorists like Paul Virilio, filmmakers like David Cronenberg ("The Palace at 4 A.M." could be titled "Videodrome"), and fiction writers like Thomas Pynchon and Philip K. Dick. However, "The Palace at 4 A.M." also alludes specifically to Alberto Giacometti and his Surrealist game boards, cages, and the like (Kessler borrows his title from one such work). The general connection to Surrealism seems clear enough: Kessler updates its strategy of pointed juxtapositions of found images and objects, which often result here, as in the best of Surrealism, in a "convulsive beauty" in which desire and death are bound up with each other. The connection to Giacometti is more precise. In his own *Palace at 4 A.M.*, with its little figures in a skeletal house that is also a nasty cage, Giacometti conjures up an obscure drama of Oedipal subject formation. In his "Palace," Kessler reflects on how we are formed as subjects today, how we are inscribed in new regimes of global entertainment and imperial politics.

In this regard, an element of paranoia—of critical paranoia—runs throughout the work, as it runs throughout Surrealism, McLuhan, Virilio, Cronenberg, Pynchon, and Dick. Dick once defined the paranoiac as the person with all the facts, while Freud viewed the paranoiac as a subject desperate to connect "all the facts," often through complex conspiracy theories, precisely because they appear so *dis*connected in the first place; for this subject, the very survival of the world seems to depend on the coherence that he or she can project on it by sheer force of interpretative will. In a further study, the Freud associate Victor Tausk focused on paranoiacs whose conspiracy theories took the form of control by "influencing machines." Clearly, Kessler plays with the tension between connection and disconnection in the world, and he, too, constructs "influencing machines" to do so (that could be another rubric for his installations). At the same time, he refuses to be at their mercy; indeed, his machines are models of how to jam, however momentarily, the image-flow of the great machines of power.

I want to conclude with one more association that, to my mind, trumps all the others. Nearly one hundred years ago, in the midst of World War I, the Dadaists in Zurich developed a strategy of mimetic exacerbation: In their Cabaret Voltaire, they took the corrupt words and images of the powers then at war and played them back as a caustic form of nonsense. "What we call dada," the great Zurich ringleader Hugo Ball wrote in 1916, "is a farce of nothingness in which all higher questions are involved; a gladiator's gesture, a play with shabby leftovers, the death warrant of posturing morality and abundance." In the midst of another war, Kessler produces a Cabaret Voltaire for our own time, its pandemonium updated to suit present conditions; and as Ball took on the role of the "magical bishop," so Kessler assumes the guise of the "global village idiot"—a shaman who seeks, perhaps impossibly, to exorcise the pandemonium of the present, pandemonium as in "abode of all demons, place of lawless violence or uproar, utter confusion."

Ludwig Seyfarth

Playground Under Surveillance

Installation Art in the Control Society Exemplified by Jon Kessler's "The Palace at 4. A.M."

"We live in an age in which highly educated professionals dedicate themselves to the task of entering the collective public mind with the object of manipulating, exploiting, and controlling."[1] This is the opening sentence of Marshall McLuhan's first masterwork of media theory, *The Mechanical Bride* (1951), a collage of chapters devoted primarily to analyzing the manipulative effect of advertising.

What McLuhan observed concerning consumer culture over fifty years ago now seems to have taken hold of the entire political and social life of America and many other nations. In the wake of the September 11 attacks, if not before, ostensibly democratic societies have increasingly become paranoid control states where as much as possible is monitored around the clock. Thanks to biometric screening, even a slightly unusual head shape can give cause for suspicion, as demonstrated recently at the railway station in Mainz, Germany. From October 2006 to January 2007, two hundred people helped to test a video surveillance system meant to identify individuals within big crowds. Experts in data security made the criticism that the system automatically recorded data on many other passing people without their consent.

In unfamiliar situations, it may even happen that one no longer recognizes oneself. Standing in front of a screen, unaware that the image is coming from a surveillance camera, you realize after a while that the person standing around, rather out of place, not cutting an especially good figure, is actually you.

Some years ago, an advertisement in Manhattan asked, "On an average day you will be captured on CCTV cameras at least a dozen times; are you dressed for it?"[2] The advertisement calls on us not to be afraid of permanent control, but to enjoy it with the same narcissism as Jennifer Ringley, a young American woman who was already streaming her private life online via webcam ten years ago, awakening curiosity above all concerning the possibility of sex scenes. Countless porn websites have since been named after "Jenny."

"The pleasure principle of the voyeur, to see everything, and the pleasure principle of the exhibitionist, to show all, have shifted from the fates of private drives to social norms,"[3] writes Peter Weibel. Narcissism, too, he adds, is changing from a psychological to a social category, a development foreseen by McLuhan as early as 1964, when he dedicated a chapter of *Understanding Media* to narcissism as "narcosis." He contradicts the traditional interpretation of Narcissus's self-infatuation and raises doubts as to whether Narcissus recognized himself in the mirror at all. McLuhan interprets narcissism in terms of media theory as man's infatuation with the devices he has created, as narcosis by the technical extensions of the body.[4]

In this reading, being in love with one's own mirror image means identifying with the mediated image of oneself to the extent of no longer being able to tell oneself apart from it. In this state, a clear view of things is just as impossible as it is in the fear-laden arousal of a state of moral outrage. Instead of not looking at all out of sheer abhorrence, one should be alert to the mechanisms that maintain a subliminal hold over us.

In *The Mechanical Bride*, McLuhan calls moral outrage a bad guide if one is not to drown helplessly in the vortex of media images. But how is one to act? McLuhan uses a literary metaphor and refers to Edgar Allan Poe's story "A Descent into the Maelstrom," in which a shipwrecked man in a seemingly hopeless situation manages to save himself from drowning. In the midst of a huge whirlpool that threatens to drag him down, he has an idea that causes him to suspect he is delirious: "I even sought amusement in speculating upon the relative velocities of their several descents toward the foam below." By behaving like the objects that are floating upward, the sailor stays alive. By analogy, in *The Mechanical Bride,* McLuhan attempts to "set the reader at the center of the revolving picture created by these affairs where he may observe the action that is in progress and in which everybody is involved."[5]

McLuhan's reading of Poe's story anticipates artistic strategies that were to develop over subsequent decades, exposing the viewer to a three-dimensional scenario that appears initially as chaotic superabundance and only becomes clearer on closer observation of details. The works in question, although very different in character, all fall within the category of "installation" in the broadest sense. Since it is more difficult here than in other fields of art today to arrive at meaningful categorizations, typologies, and genre definitions, it seems appropriate to programmatically refrain from doing so, as Juliane Rebentisch does in her study that approaches the theme of the installation from the point of view of philosophical aesthetics.[6] This notwithstanding, I would like to outline some basic tendencies in order to give a more precise idea of the specific qualities of Jon Kessler's installation art.

The tendency toward the chaotically overfilled installation, an insatiable accumulation of materials, may appear hypertrophic and megalomaniac. Listing them more or less chronologically from the 1960s until the present, I am thinking here of artists like Edward Kienholz, Paul Thek, Anna Oppermann, Paul McCarthy, Mike Kelley, Jason Rhoades, John Bock, and Jonathan Meese.

If I do not include Jon Kessler and "The Palace at 4 A.M."[7] in this "category," then it is because in spite of their size and fullness, there is a different tendency or tradition at play in his works, one of scenarios that are sparser and more clearly structured. A typical feature of this installation category is an often-disconcerting use of security cameras, which we also find in Kessler's work—about which more later.

Installations occupy space, giving them a theatrical or stagelike quality, even if no dramatic action in the form of a performance or happening takes place. As such, an installation is also a kind of picture rendered in three dimensions that depends on the actions of its viewers to set it in motion. This is true of "The Palace at 4 A.M." in a particularly subtle sense, thanks to the presence of a "second level" where that which appears to move (i.e., that which we see on the innumerable screens) is also frozen at a standstill. The influx of images is strictly limited in two ways. First, no prerecorded images are shown (with the exception of a computer-simulated satellite picture and map of Iraq, a sequence that ends in a satellite image of Hamburg-Harburg, suggesting a bomber pilot's view of the exhibition venue. A corresponding video was also featured when the installation was first shown at P.S.1 in New York). And second, the screens only show what is to be found within the installation (including individual views out of the windows). In spite of its technical sophistication, the installation is in fact an apparatus that does not really "work."

This can be compared with the installations of the Belgian artist Guillaume Bijl, who in 1979 began to reconstruct consumer venues and other locations of life within society—but purely as exhibits. In Bijl's driving school, you cannot take a driving test; you cannot shop at his supermarket, you cannot bid for anything at his auction house, and no quiz show is being filmed at his television studio.[8]

For all its supposed realism, Bijl's *TV Quiz Decor*, 1993, is a dummy, a pure facade. Looking behind the scenes of the brightly glittering scenario, one does not find any kind of studio capable of performing any kind of broadcast. Although in Kessler's work the technical installation works, it only registers what is inside the installation. Whereas in Bijl's installations there is no behind-the-scenes, Kessler realizes the "backstage" but leaves nothing out front that could be meaningfully processed.

While Bijl presents the settings of consumer society as nonfunctioning still lifes, Kessler does a similar thing with reporting on war and terrorism in the media. He leads the viewer into a gigantic homemade playground of war with soldier figures; punched and bent sheet metal; large-format photographs mounted on plywood; postcards of the World Trade Center rotating like a paternoster lift; apparatuses performing various circular, rotary, or jerky movements to which pictures, various objects, or cameras are fastened; endless cable connections, often hung right across the space; and innumerable screens, suspended, standing on podiums, or assembled into walls.

Kessler displays the entire technical apparatus to us, frustrating any kind of voyeurism by showing that, for the voyeuristic gaze at least, there is nothing to see. We are allowed to walk in and through anywhere, to crawl through narrow openings, even through the cutout crotch of a massively enlarged porn photo, but we will never get through to what lies behind, to the inside of the labyrinth, because there is no inside. We already are "inside." The cables and devices are like the exposed bundles of nerve fibers of the worldwide electronic brain that McLuhan has long since declared us to be parts of.

The unsettling and reversing of relations between interior and exterior is also characteristic of installations created around 1970 by artists whose work addressed the relationship between architecture and surveillance, especially Bruce Nauman and Dan Graham. One is filmed, sees oneself from behind or with a time lapse somewhere else in the exhibition. Nauman is interested in the medium of video "as an electronic mirror and surveillance device."[9]

Environments whose construction appears simple develop complex relationships between seeing/not seen and observed/hidden. In his current works, Graham still uses the two-way mirrors that he first featured in his installation *Two Viewing Rooms* in 1975. The mirror separates two adjacent box rooms: One room is dark, the other light. In the light room, as the visitors come in, they see their own image in real time on a screen positioned in front of the mirror. In the dark room, other visitors can observe those in the light room, who do not notice they are being watched until they realize that the camera filming them must be behind the mirror.

Graham has also written theoretical texts dealing with the social and political function of surveillance systems.[10] A number of artists have developed the Nauman-Graham paradigm in various directions. Peter Weibel's *Beobachtung der Beobachtung: Unsicherheit* (Observation of Observation: Insecurity), 1973, was an early European take on Nauman's closed-circuit monitor setups. Since the 1980s, Julia Scher's work has dealt consistently and in immediately recognizable form with the issue of surveillance. Other artists and filmmakers analyze less the technical equipment than the automatically recorded image material. Michael Klier's film *Der Riese* (The Giant), 1984, consists almost entirely of footage recorded by surveillance equipment in major German cities. For *Suicide Box*, 1996, the Bureau of Inverse Technology edited together images from a camera that was aimed toward the Golden Gate Bridge in San Francisco for one hundred days, also recording the suicides regularly committed there.[11]

Crucial inspiration on the theme of surveillance, on links between power relations and lines of sight, came from the theoretical writings of Michel Foucault. In *Discipline and Punish* (1975), Foucault analyzed Jeremy Bentham's panoptical prison concept from the early nineteenth century. From the center of the circular structure, the guards were able to monitor the prisoners in the surrounding cells, while themselves remaining unseen by the inmates. In 1979, Dutch architect Rem Koolhaas's entry for a competition to redesign Arnheim jail proposed a reversal of the one-way lines of sight, with the guards sitting in glass-walled rooms. It goes without saying that Koolhaas's project was not realized. Although it was conceived of as "real" architecture, in social terms it was so utopian that it took on a status between architecture and art, coming close to some of Graham's ideas.[12]

"Many of the surveillance systems we have today are based on Bentham's principle—especially where the devices are left switched off or are just dummies anyway, containing no film or video tape or transmission cables, but still fulfill their function, then they recall Bentham's Panopticon,"[13] writes Thomas Y. Levin, curator of the "Ctrl [Space]" exhibition at ZKM Karlsruhe (2002). The only direct comparison, however, is the uncertainty over whether one is actually being watched. Streets, shops, garages, banks, stations, and airports—the indoor and outdoor urban spaces watched over by security cameras—are not prisonlike "sites of confinement." In Foucault's analysis, these sites of confinement are characteristic of the disciplinary societies that existed from the early nineteenth century until World War II. They were marked by established institutions—the prison, the hospital, the factory, the school, the family—which regulated both normal social life and temporary or permanent exclusion from it, for example by means of prison sentences. But even the lives of individuals who were not incarcerated passed through a series of sites of confinement: school, university, long-term job, etc.

Although these sites of confinement have not disappeared, the disciplinary societies have increasingly transformed themselves into "control societies"—a process described by Gilles Deleuze in his development of Foucault's theory of power.

In the control society, the role of architecture in constituting sites of confinement breaks down. Instead of disciplining people by subjecting them to rules in spatially confined settings, controls take place everywhere and all the time. Surveillance is no longer something exercised tangibly on the—confined—body. At the same time, one never knows where one really stands in life. One is expected to develop continually, keep learning, change jobs—to be "flexible," as discussed by the American sociologist Richard Sennett in *The Corrosion of Character* (1998): Factories with rigid hierarchies have been replaced by "enterprises," owners have been replaced by a succession of managers, and jobs may now be lost due to bad speculation on the part of the shareholders.

In his "Postscript on Control Societies" (1990),[14] Deleuze cites the use of electronic tagging as an alternative to conventional custody as a metaphor for the control society. The offender does not have to stay in jail, but he is also not really free in his movements, since his location can always be checked. The inventor of electronically controlled house arrest, a magistrate from New Mexico, has explained that he got the idea from a 1979 Spiderman comic.[15]

Unlike the early installations by Graham and Nauman, in which one is subjected to a kind of test setup in more or less closed spaces, in the case of Jon Kessler, the viewer is more like someone tagged with an electronic shackle. You can walk around as you please, but the control system always knows where you are. And when we notice a familiar face from our circle of friends on one of the screens, we may be just as surprised as if we had seen it on television. We just happened not to know that this friend was also visiting the exhibition at the same time as us.

Real-time transmission makes us sure the friend really is present in the space. This authenticity effect is seen by Thomas Y. Levin as the main reason for the ever-increasing interest in security cameras and real-time broadcast: "The 'semblance of reality' . . . which in the past was guaranteed by the indexicality of photography is now achieved by the rhetorical power of so-called 'real time.'"[16]

Jon Kessler takes this semblance of reality and renders it absurd. What we see on the screens are newspaper photos, collages, toy soldiers, punched and shaped pieces of sheet metal, "Fuck You" written in black paint, lines drawn on a rotating drum (quoting Duchamp's chocolate grinder), puppets hanging on the wall, light refracted through colored transparencies—pictures and three-dimensional arrangements possessing no value in terms of a reality effect. Kessler thus takes literally postmodern image theory, which, according to Peter Weibel, "does not begin with an observation of the world, but, rather, with an observation of the image."[17]

The insight that art, too, is a contemplation of images and not of the world also has a pleasant influence on the way Jon Kessler deals with September 11 and ensuing changes in the political climate. He steers well away from the idea that an artwork could compete with acts of terrorism and their real effects—the idea behind the unfortunate comments by Karlheinz Stockhausen, who, in a press conference a few days after the attacks, called them the greatest artwork ever.[18]

But maybe his argumentation was simply out-of-date, repeating thoughts aired more than two centuries ago by Edmund Burke, who, in his *Philosophical Enquiry* (1757), wrote: "Chuse a day on which to represent the most sublime and affecting tragedy we have; appoint the most favourite actors; spare no cost upon the scenes and decorations, unite the greatest efforts of poetry, painting, and music; and when you have collected your audience, just at the moment when their minds are erect with expectation, let it be reported that a state criminal of high rank is on the point of being executed in the adjoining square; in a moment the emptiness of the theatre would demonstrate the comparative weakness of the imitative arts, and proclaim the triumph of the real sympathy."[19]

The idea that art should be seen not as something that competes with reality but as its reflection is already present in Kant's thinking on the sublime. By relating the property of sublimity not to physical objects but to the subject's cognitive faculty, Kant presents a way out of the dead end into which Burke's argumentation automatically leads. But even today, Stockhausen is not alone in attesting to the relative weakness of artistic fictions. One year later, Damien Hirst looked back with admiration at the real power of the terrorist act: "The artist Damien Hirst said last night [in an interview with BBC Online News] he believed the terrorists responsible for the September 11 attacks 'need congratulating' because they achieved 'something which nobody would ever have thought possible' on an artistic level. Hirst, who is no stranger to controversy, said many people would 'shy away' from looking at the event as art but he believed the World Trade Center attack was 'kind of like an artwork in its own right.'"[20]

The question of whether or not acts of terrorism and wars are artworks is not even raised by Kessler, as his installation competes neither with actual reality nor with the global power of visual media. As they rattle, gyrate, wobble, or clatter like artillery fire, his apparatuses can hardly constitute an attempt to compete with the technological and strategic might of those responsible for the global distribution of images of 9/11 or the Iraq war. While the freewheeling motion of Jean Tinguely's motorized iron sculptures playfully ridicules the functionalism of the machine age as it comes to a close, Kessler's ramshackle theatrical arrangement can be seen as a grotesque caricature that parodies the global diffusion of stereotypical media images in the form of a gigantic kinetic sculpture.

Kessler's sculptures and objects from the 1980s and early 1990s continued in the tradition of kinetic machines. It is only in recent years that Kessler has also begun to focus on video technology, which he uses mainly, as in "The Palace at 4 A.M.," in the form of small surveillance cameras integrated into his objects and installations. An intelligent game of make-believe is also played out here. Kessler proceeds as if the development of image-producing media to the present day had taken place as it has done—except that there is still no possibility of recording and storing pictures. This leaves us with one foot in the eighteenth century, when Lessing discussed the problem of motion using the example of a marble sculpture.

But even at that time, there were numerous mechanical apparatuses and toys that generated or simulated motion and also made use of the afterimage effect later exploited by film. The richness of this pre-cinematographic world with its links to the cabinet of curiosities is evident in the well-documented collection of the German filmmaker Werner Nekes, which includes anamorphic pictures, distorting mirrors, peep shows, shadow theaters, and optical conundrums.[21]

Many of the apparatuses in "The Palace at 4 A.M." recall the simple construction and the often-sophisticated play with surfaces featured in these pioneering developments, which, although still unable to record them, made images move.

In this light, one might also say that Kessler's installation equips the world of mechanical picture machines and pre-digital kinetic sculptures with modern security cameras. But it would not be possible to entirely forego chemical and digital recording. Otherwise, the reproductions hung on the walls, mounted on stands, or revolving mechanically in circles could not exist.

Moving to another, slightly different level of description, one might ask whether Kessler is not also acting as if the camera obscura model of perception were still current and had not been, as Jonathan Crary demonstrates in *Techniques of the Observer*, replaced during the last century by a neuronal model. The fact that science no longer described seeing as light falling into the eye, but as the stimulation of nerve cells, led in the long term to the development of "techniques that are relocating vision to a plane severed from a human observer. . . . Most of the historically important functions of the human eye are being supplanted by practices in which visual images no longer have any reference to the position of an observer in a 'real,' optically perceived world." This "relentless abstraction of the visual"[22] is described by Paul Virilio, whom Kessler cites explicitly, as a "vision without a gaze."[23] Kessler denies this separation: Every picture that appears on one of the screens reveals itself to our gaze, which identifies it and thus locates it within the space. As in the case of a sculpture, the viewer perceives the full complexity of the installation by moving around it, seeing it increasingly clearly, making a mental image of its overall form.

The reception of the installation, then, is also an imaginary drawing in space, and in this sense, Kessler seems to have borrowed more than just the title of Alberto Giacometti's table sculpture *The Palace at 4 A.M.*, 1932. Giacometti's palace, a structure consisting mainly of thin struts, looks like a drawing materialized in three dimensions.

In spite of the muddle, the spatial relations in Kessler's work always add up. The arrangement of the various individual elements is dictated by his sculptural approach, which constitutes a fundamental difference between his approach and that taken by many other artists who see their installations and performances not in terms of sculpture, but as a kind of three-dimensional painting. Kessler's work also lacks any performative element, except, of course, that visitors might deliberately act out performances that could be watched somewhere on the screens by other visitors. Ultimately, each individual who walks through the installation sees something different, making Kessler's stage-like arrangement similar to the recent work of Christoph Schlingensief, the German director and actionist who directed *Parsifal* in Bayreuth for the past two years. Lately he has produced immersive installations such as *Church of Fear* at the Venice Biennale and his "Animatographs," such as *Kaprow City* at Berlin's Volksbühne Theatre. The concept of the "Animatograph" is a revolving stage consisting of two counter-rotating discs on which are positioned many separate rooms of different sizes where parallel plot strands are acted out by the cast of a film that is produced and edited live. In *Kaprow City*, one section of the audience is seated on the stage, another off the stage behind the curtain. Most sit in the auditorium and are separated from the action onstage almost the entire time by a projection screen. But they see direct transmissions of individual scenes in poor quality, as if recorded by a webcam.

Instead of bringing the different viewing levels together, the confused spectacle surrounding Jenny Elvers-Elvertshagen's performance as Lady Di, a homage to the inventor of the happening, Allan Kaprow, tends instead toward entropic decay, also—at least in the performance I saw—in terms of the way the live action increasingly stagnates and breaks down into discrete moments. And the use of the camera corresponds to the pornographic character of video surveillance predicted by Warhol, who said people would use video cameras mainly to make porn films and spy on their neighbors.

Kessler, too, has the aggressively pornographic aspect of universal media visibility in his sights, but the way he highlights things has more in common with the nonchalant way staff at Cambridge University realized the first online broadcast via webcam in 1993. It showed the coffee machine in the computer lab, and it was "on air" until 2001. Did anyone keep watching the whole time?

Anyone who keeps watching with Kessler will keep discovering new things, only to realize in the end that everything stays the way it is. "The Palace at 4 A.M." is one of the subtlest media and video sculptures made to date. Like Nam June Paik's video sculptures, one could see it as an allegory of our media age. But Kessler avoids direct objective and symbolic associations such as the robot-like figures, crosses, stars, and gates formed by Paik out of dozens of monitors. Kessler neither relies on the easily identified pictoriality of such a "narrative-objective overall structure"[24] nor uses the physical heaviness of sculptural materials to symbolically crush the fragility of technical devices, as Wolf Vostell did, for example, with his televisions and cars set in concrete. Rather than immuring us in a *Depression Endogène* (as Vostell titled one of his television sculptures in 1984),[25] Kessler leads us into a seemingly open system where, like Edgar Allan Poe's sailor, we must first locate ourselves; and images are the orientation.

1. Marshall McLuhan, *The Mechanical Bride: Folklore of Industrial Man*, Gingko Press 2002 (1951).
2. Quoted in: Thomas Y. Levin, "Rhetoric of the Temporal Index: Surveillant Narration and the Cinema of 'Real Time,'" in: *Ctrl [Space]: Rhetorics of Surveillance from Bentham to Big Brother*, exhibition catalogue, ZKM Karlsruhe, Karlsruhe/Cambridge, Mass. 2002, pp. 578–593.
3. Peter Weibel, "Pleasure and the Panoptic Principle," in: *Ctrl [Space]*, op. cit., pp. 207–223.
4. Marshall McLuhan, *Understanding Media: The Extensions of Man*, paperback, London 1967, p. 51 ff.
5. McLuhan, *The Mechanical Bride*, op. cit.
6. Juliane Rebentisch, *Ästhetik der Installation*, Frankfurt 2003.
7. "The Palace at 4 A.M." is based in part on the installation "Global Village Idiot" (2004, Deitch Projects, New York). The title, a reference to McLuhan's concept of the global village, is also the title of a collection of *Guardian* columns by the English best-selling author John O'Farrell, with critical and satirical political commentaries, including some on the Iraq war.

8. The most extensive analyses of Bijl's oeuvre are: Frank van de Veire, "The Funerary Scene of the Object: On the Installations by Guillaume Bijl," in: *Guillaume Bijl*, exhibition catalogue, Museum van Hedendaagse Kunst, Antwerp, Kunstverein Hannover, Freiburger Kunstverein, et al. 1996, pp. 17–39, and the chapter on Bijl in: Stefan Römer, *Künstlerische Strategien des Fake: Kritik von Original und Fälschung*, Cologne 2001, pp. 215–239. However, by concentrating on the conceptual side of the work and on Bijl's reception by art critics, Römer doesn't devote sufficient time to strategic illusionism, which also manifests itself in the way Bijl understands the real objects he uses as sculptures.

9. Barbara Engelbach in: *Bruce Nauman: Versuchsanordnungen. Werke 1965–1994*, exhibition catalogue, Hamburger Kunsthalle 1998, p. 37.

10. See "Video in Relation to Architecture" and Graham's texts on individual surveillance installations in: *Dan Graham, Two-Way Mirror Power: Selected Writings by Dan Graham on His Art*, Cambridge, Mass./London 1999.

11. The best overview of artistic approaches to the issue of surveillance is: *Ctrl [Space]: Rhetorics of Surveillance from Bentham to Big Brother*, exhibition catalogue, ZKM Karlsruhe, Karlsruhe/Cambridge, Mass. 2002.

12. Rainer Metzger discusses Koolhaas's project in relation to Dan Graham in his book: *Kunst in der Postmoderne—Dan Graham*, Cologne 1996, p. 149.

13. Thomas Y. Levin, "Die Rhetorik der Überwachung: Angst vor Beobachtung in den zeitgenössischen Medien," in: *7 Hügel. Bilder und Zeichen des 21. Jahrhunderts*, exhibition catalogue, Berliner Festspiele, Berlin 2000, volume 6: *Zivilisation*, pp. 49–61.

14. Gilles Deleuze, "Postscript on Control Societies," in: *Negotiations 1972–1990*, Columbia University Press 1995, pp. 177–182.

15. Cf. Levin, *Rhetorics . . .* op. cit., p. 58.

16. Levin, *Die Rhetorik . . .* op cit., p. 56.

17. Weibel, op. cit., p. 210.

18. When asked whether there was not a difference between art and crime, he did admit that he considered the terrorists criminals, "because people did not consent. They didn't choose to attend the concert," but at the time, most people saw his position as nothing but inappropriate cynicism. Transcript of the press conference of September 16, 2001: *MusikTexte* 91, November 1991, pp. 69–77 (online: http://stockhausen.org/hamburg.pdf).

19. Edmund Burke, *Philosophical Enquiry into the Origin of Our Ideas of the Sublime and Beautiful*, Oxford University Press World Classics 1998, p. 43.

20. *The Guardian*, September 11, 2002 (online: http://www.guardian.co.uk/september11/oneyearon/story/0,12361,790059,00.html).

21. *Ich sehe was, was du nicht siehst! Sehmaschinen und Bilderwelten: Die Sammlung Werner Nekes*, exhibition catalogue, Museum Ludwig, Cologne, Göttingen 2002.

22. Jonathan Crary, *Techniques of the Observer: On Vision and Modernity in the Nineteenth Century*, Cambridge/London 1992, pp. 1–2.

23. For example in: *The Vision Machine*, London 1994.

24. Konstanze Thümmel in a catalogue piece on Paik's *Passage*, 1986, in: *Kunst der Gegenwart*, inaugural catalogue, Museum für Neue Kunst/ZKM Karlsruhe, Munich/New York 1997, p. 202.

25. Heinrich Klotz wrote that "Vostell's *Depression Endogène* is an illness which offers pictures but which nearly chokes them in concrete or, worse still, pleasantly trivializes them" (ibid., p. 276).

General Arthur Prinzhorn Ethan Prinzhorn & Jon Kessler

Conversation

This interview took place in Yonkers, New York, on June 5, 2006. General Arthur Prinzhorn is a highly decorated, four-star general who retired from the U.S. Army in 2003. His youngest son, Ethan, will be receiving his M.A. in curatorial studies from Bard in 2007. The Prinzhorns were the next-door neighbors of the Kesslers in Yonkers.

General: Let's get right down to it, shall we? Is your show critical of the current war?

Jon: The show is certainly a response to our current situation. Personally, I am highly critical of the Bush administration's disregard for our Constitution and the White House's kowtowing to the interests of right-wing groups. With an acquiescent media, our government misled us into a war in Iraq, and it's clear that the same thing is happening now with Iran. But the show is less a polemic against Bush than my personal coming to grips with the rage and alienation that I've been feeling of late. How about you, General? You have certainly been very outspoken recently about the Iraq war and certain members of the Bush administration, especially Donald Rumsfeld.

General: As a general, you have a responsibility to the men and women that serve under you, as well as to the civilians that you are sworn to protect. I spoke out against Donald Rumsfeld because he has and continues to display a complete disregard for the men and women of the military and a total contempt for the American people. The way he wants to fight this war is untenable—we're just killing innocent people as collateral damage for a failed vision of a "new" Middle East. While I support the spread of democracy in the region, and believe that it is incumbent that the region experience political change, the Bush administration is going about it the wrong way. They may very well drag us into a nuclear war with Iran where there are no winners, only losers. September 11 should have been a wake-up call for America to open its eyes and see how our foreign policy and dubious attempts to protect freedom are seen by the rest of the world. But except for a short period of flag-waving, we went back to sleep. I say this as Congress spends the next three weeks debating same-sex marriage—my GOD!

Ethan: But Dad, why were you so critical when we marched in protest of the war? Isn't that an expression of the kind of democracy that you're trying to spread in the Middle East?

General: I never critiqued your right to protest, I just knew that it was useless. I had never witnessed an administration so determined to go to war despite the worldwide mobilization against it.

Ethan: So I think we're all in agreement that we feel lied to—Jon, how does your show react to this perceived injustice?

Jon: Five years ago, if you had told me that I'd be making work that addressed this subject matter, I wouldn't have believed you. I feel like Michael Douglas in *Falling Down* or Peter Fonda in *The Wrong Man*—you can only be pushed so far before there's a reaction. In 2001, I was working on a series of mobiles that incorporated body casts of my wife, my daughter, and myself. After 9/11, I couldn't get the image of what the terrorists saw from the cockpit out of my head, so I decided to re-create it. That impulse led to *One Hour Photo*, and since then the work has taken on this more politicized direction.

Ethan: So there was an explicit change in your work that was sparked by 9/11?

Jon: The change in the work was initiated by the subject matter, but it was propelled forward by the introduction of video into the sculptures. This made the work come alive again. In 2004, when I did the show "Global Village Idiot," I hadn't shown my work in ten years. I had lost my voice and didn't really know what to say anymore. The introduction of video into the work in 2001 catalyzed the breakthrough in the work that I was looking for. People have commented on a return to a looser, unfinished attitude in the new work. Throughout the '90s, the work relied heavily on production, and I was spending countless hours doing things in the studio that I didn't like, like polishing aluminum. In the recent work, when I paired the machines to video monitors, it allowed me to be more direct in constructing the mechanisms. At the point when they achieve the desired effect for the camera, I leave them alone.

Ethan: In the '80s, you became known for doing specifically sculptural pieces that embraced elements of kitsch and found objects—how do you see this current show as an extension of the ideas put forth back then?

Jon: I was always interested in getting people to look behind the curtain. Getting them to become active viewers, to investigate the mechanism, to suspend their disbelief, and, finally, to have an experience with the objects that I was presenting, even if many of those objects originated as kitsch. The very early work played with pictorial space by creating a duality between the mechanism and the screens that was hopefully more than the sum of its parts. In many ways, this recent video work is a return to this duality. It's funny that you say I became known in the '80s. The changes in the work were facilitated by the fact that I felt completely free to reinvent myself. I had lost or left all of my galleries, and there was little interest in the work. When *Artforum* published the double issue on the '80s and there was no mention of me at all, I really knew that no one was watching me.

Ethan: I would also say that there is a related theme in both your early works and your current show, which concerns the fetish. The early work, it seems, plays with this notion of the fetish in relation to objects—the complex power that we lend kitsch, for instance. Similarly, in this current show, you refer to the power that we imbue in the event—the fetish of 9/11. Does this resonate? Are you ever afraid of falling into that trap of fetishizing 9/11 as so many have, politicians as well as artists?

General: I'm going to go one step further than my son and say that your show was exploiting the events of 9/11!

Jon: First to Ethan's point, the works from the '80s did address commodity fetish, although this was never foregrounded in my work the way it was in my contemporaries such as Jeff [Koons] or Haim [Steinbach]. In the Asian-inspired works, it was more of an attempt to fetishize the culture—turning exoticism and otherness into a commodity. As for fetishizing and exploiting the events of 9/11, there is no overestimating the harmful effect that Al Qaeda's ability to stage a truly murderous image-event had on the control of image production in our culture. September 11 is a constant reminder of America's vulnerability and proof that we no longer have a monopoly on big violence.

General: I don't know if it was your intention, but the sounds produced by the mechanisms in the show sounded like artillery fire. I felt like I was back in the trenches.

Jon: Throughout the summer, I had only one sniper mechanism running in the studio. "Snipers" are the mechanisms that move the cameras in quarter turns by use of Geneva mechanisms. Honestly, I hadn't predicted the rhythm of the sound of seven of them functioning at the same time. I thought it sounded like a chain gang, but many, like you, were reminded of the sound of artillery—lock and load. I found the immensity of the P.S.1 space daunting, so I chose to sculpt the space with sight lines, targets, and zones of vision instead of physical mass. The snipers also made the connection between surveillance and the new eye of totalitarian war—smart bombs.

Ethan: One of the many contradictions that gave the installation its tenor was the precision of the mechanism's control over the mediated "space" in contrast to the looseness of the installation, with the hanging wires and the way that the machines were constructed.

Jon: Control over mediated space is one of the implied subtexts of the show, especially through the use of technology. So in some ways, the show is commenting on, as well as an example of, this. It drives this principle at the same time as it undermines it.

Ethan: In 1954, Michel Carrouges found a structural similarity between Duchamp's *Large Glass* and the torture device described in Franz Kafka's short story "In the Penal Colony." His insight was that both were closed systems with one part imposing on the other. This reminds me in several ways of your installation and your evocation of Abu Ghraib.

Jon: The installation does become a series of closed circuits that emanate, affect, and impose on each other. Abu Ghraib was suggested in *Theater of Ideas* with the GI Joe who is standing on a box with his pants pulled down and wearing a dunce cap. Surrounding him are more toy soldiers hanging from the structure of the sculpture; these are meant to evoke the images of the U.S. civilians who were burned and hung from the bridge. Surrounding them are embellished images of Kabuki, a form of highly stylized theater from Japan, and they suggest a kind of scary "other" that a soldier from Kansas might experience out on night patrol. Finally, the red splatter is meant to frame and encapsulate the order, control, and sensuousness of violence much in the way Kurosawa uses blood in his films.

General: There is something sensuous about violence. For many soldiers, fighting becomes as much an addiction as a duty. It's funny, because walking through your show oftentimes I wasn't quite sure if I liked your work because the images were attractive or because they were so gruesome. Is there a history of artists reacting to war?

Jon: For me, Dada is the most poignant example—they were responding to the mechanized bloodbath of World War I. But the title "The Palace at 4 A.M." comes from a great Surrealist work . . .

General: What does the title refer to?

Jon: It refers to a Giacometti sculpture of the same name in MoMA. I've always loved the piece for its sense of staging, a theatrical mise-en-scène that feels like a model for a stage set about a dream sequence. When I read reports of our troops stationed in Saddam's palaces and heard about Abu Ghraib, I thought "The Palace at 4 A.M." was an appropriate title for the entire exhibition; it evokes the insanity that happens at that hour, when no one is watching.

General: Did the show change a lot when you mounted it in Hamburg? It seemed so specific to the P.S.1 space. How is the show in Hamburg different?

Jon: The interesting thing about the show is how flexible it is. Since the snipers are based on zoom lenses, it doesn't matter if the target is five feet away or a hundred; the show could expand and contract accordingly. In New York, the show was mounted in that giant room, which acted metaphorically as the grand public ballroom of the palace. The smaller rooms adjoining this were more private chambers, in which I placed many of the figurative works. In Hamburg, I used the billboards to create a series of rooms that allows the viewer to come upon the show's ideas in a slower fashion. When they view someone on a monitor, they can't just turn around and see them, since they will most likely be in another room. I cut holes in the walls and mirrors, and the billboards have been refigured to create more of a labyrinthine trip through the looking glass; sections of the show are lit with flashlights mounted to the cameras and acting like searchlights. *Shock and Awe*, a piece in which a photograph of Baghdad being bombed is cut and taped to the window in such a way that it covers the actual sky and becomes part of a composite image, turning the outside world into a prop, took on an odd and unexpected dimension in the Hamburg installation. A camera, trained on the street, happens to photograph the neighborhood where Mohamed Atta had lived while he planned the destruction of the Twin Towers. The idea of this show began as a reaction to the toppling of the World Trade Center. In a haunting and bizarre bit of chance, the Hamburg installation returns to the place where 9/11 was masterminded. There are also some new pieces for Germany, such as a sniper that uses an airplane window from my flight over . . .

General: You mean a picture of the window on your airplane?

Jon: No, no . . . the actual window.

General: Jon, you stole an airplane window?!

Jon: Well, I guess I did. Hey, the plane was literally falling apart. When we landed, the window had come loose, so I stuffed it in my bag. It's inset into a wall with a view of a brick wall through it. There's also a new piece at the entrance that has a camera inserting itself into a silicone vagina—this piece creates a double insertion, one with a camera entering a vagina that brings you into the mediated world telegenically, and then the billboard of the woman's crotch that you physically walk through to enter the show. It immediately sets up the symbiotic relationship between the camera and the viewer, which continues inside as spectator and performer, voyeur and exhibitionist.

General: Well, I can't say that the missus liked walking through the vagina to get into your show. It seemed a little blatant and, according to her, a poor decorating choice.

Jon: I was really thinking about the relationship between pornography and war that surfaced when the Iraq invasion began and the only place to see uncensored footage of battles was on amateur porn sites. These sites began as a place to post nude photos of girlfriends and wives but then became a place for U.S. troops to post videos of gruesome war scenes so that their family and friends could have a more accurate idea of what was really going on in Iraq. The viewer enters the show through a vagina, which belongs to a wall-size photo of a naked woman from behind. The image comes from an X-rated magazine, but all the sexually explicit details have been removed. I wanted the viewer to be reborn into the show; however, I didn't want this to be a clean or simple birth, but one complicated by the media's infatuation with disaster and titillation. These image-collages are also mimetic, commenting on the media's own desire to excite and unnerve.

Ethan: True. Throughout the show, you use still images that mirror the titillation and violence of the actual show, so you constantly have this double experience of understanding this chaos interactively and through the still. Can you talk a bit more about your use of appropriated images and source photos? For example, what about the piece *The Office*?

Jon: I began with four coffee-table books that I bought at the Strand bookstore in New York—all of them were tributes to the efficiency of American power and might. These books were the readymades, and I used them as a starting point to build my own narratives. The pages got ripped out, cut into, painted on, and collaged together to form new stories. This propelled the work forward, and I built the machines to drive the narratives. In New York and Hamburg, *The Office* is the first piece that you see, and it presented the idea of the corporate lull, an antidepressant-laced monotony, the incessant calm that is only a prelude to the shit storm that you were about to experience through the encroaching entryway. I also think of it as a recruiting office; traps that are set in small towns to enlist undereducated, undervalued human flesh and bones that have become yet another part of Cheney's collateral damage in Iraq.

Ethan: And how do the *Swans* figure into this ennui or sedation?

Jon: The *Swans* refer to extreme-makeover reality shows such as *The Swan* and *Nip/Tuck*. I took images from popular women's magazines and made cuts in them to reconfigure their shape and image. The strategy is tied to Dada practitioners of photo collage such as Hannah Hoch and John Heartfield, but sculpturally they follow a line that began with Picasso's Cubist sheet-metal guitar. The viewer completes the representation of the women, which are mounted on walls, as they approach the pictures; a surveillance camera captures the composite image on a screen. It is sort of a riff on that line from the movie *Jerry Maguire*, "You complete me"; in these pieces, the viewer literally does. The *Swans* refer to the fashion, commodity, and celebrity culture that dominate our lives. They are a way of taking back some power by subverting the images that control us, making these images our own. I made most of the *Swans* in a week; one led to the next and there were some violent ones. I took a sledgehammer to Winona Ryder's face. But most of the pieces were produced with nip and tuck procedures—after the photograph was mounted to aluminum, it was cut with a jigsaw. I must admit, I was in a weird violent zone last summer as I was creating the P.S.1 show. I had to work fast, and it didn't allow me time to doubt myself. Only having four months to produce the show was probably a positive thing.

Ethan: Jon, but I find it a bit disconcerting that all the images that you tore apart are of women. Is there a reason for this?

Jon: I tore apart men's faces elsewhere in the exhibition, but for different reasons. The *Swans* are all women because we live in a culture where women's bodies are objectified, and with television shows like *The Swan* and *Nip/Tuck*, these pieces are a kind of extension of what's already happening. Women's faces and bodies are distorted in ways in which the artificial becomes the ideal. Individual traits are replaced by shapes and textures never found in nature; bumpy noses, crooked teeth, frizzy hair, little breasts are erased; and real people are turned into freakish fakes. Lips and breasts and brows are so incredibly distorted but have become so ubiquitous that they're beginning to look almost normal, almost natural. I've always liked playing with that in my work—artificial nature.

General: Who is the hairy guy? He's definitely gone AWOL.

Jon: Yes, General, he has. He was originally made for *Party Crasher*. I used an English collectible figure and customized him with horsehair. In fact, he's an Arab character. My initial instinct was to make a self-portrait: an aging, suntanned hippie; me, if I had made a few different life choices. But when he was finished in the studio, people would come by and ask me who the homeless guy was, or the terrorist. So he became my hippie/homeless/terrorist dude, a kind of floating signifier. For "The Palace," he returned as the central character in *Evolution*. Using two different Arab heads, I customized the readymade to make four characters. The transition has him waking up in the burning embers of the WTC, on a blog site kissing the troops in Iraq, as a four-star general, and, finally, after receiving a Botox treatment, he becomes a Republican Senator.

General: Are you trying to say that these types are one and the same?

Jon: They all occupy the same world stage at the moment, and they push up against and react to each other.

Ethan: Can you talk about why you used that one piece of prerecorded footage from Google Earth that showed P.S.1 and Phoenix getting blown up?

Jon: Like the cameras trained on the street, I wanted to further complicate the proposition of the exhibition by adding footage that I might have made in my own low-tech way using cameras, photographs, and machines. Besides, Google Earth is the most amazing surveillance source available for artists to steal from.

Ethan: Do you think of your work as interactive art?

Jon: My work has included the viewer for many years, so in that sense it's interactive. *Isolated Masses* from 1985 had a heater that slowly warmed the viewer if they got close to the work. *Path of a Carp* from 1987 had an electric eye and voice chip that welcomed the viewer in

Japanese. In my new work, everything changed when I removed the background in *Party Crasher* and the hairy dude occupied the same space as the viewer. This premise of including the viewer continued in *Heaven's Gate*, *Gisele and the Cinopticon*, and exploded in the "The Palace at 4 A.M." The viewer certainly interacts with my show whether they want to or not by constantly entering the work—completing and disrupting the camera's sight lines.

Ethan: How much of the show was conceived in your mind beforehand?

Jon: The original pitch to P.S.1 was nothing like how it turned out. The ideas for the show snowballed as I began working with images that I tore from the books and spending more time at P.S.1 over the summer. The show grew organically and changed direction as I progressed, much like the way my individual sculptures evolve. In the end, "The Palace at 4 A.M." turned out to be much more image-based than my previous show, "Global Village Idiot."

General: I remember that your dad had a workshop in the basement of your house, but I don't remember you spending much time there. Were you trained in engineering?

Ethan: How would you know if he spent time there? You were never around!

Jon: My dad was a bricoleur and my brother is an engineer, so the genes are in the family. I've picked things up along the way, and the mechanisms are based on trial and error. I had the idea for a mechanism that would have intermittent motion and turn a camera 90 degrees. I worked on this for weeks, and when I finally succeeded, someone came to my studio and informed me that it was a Geneva mechanism, invented centuries ago for clock mechanisms.

Ethan: You always talk a lot about play in your work. How do you build play into the process?

Jon: This is difficult. Making work that has to function involves a lot of methodical labor and engineering, so I have to create spaces in the pieces that allow a more intuitive, faster way of working. Almost all of my sculptures begin with me sitting on the floor in my studio playing with a video camera and found objects.

Ethan: It's obvious how the airplane images work in your installation. Why did you include the cutouts of Hummers?

Jon: I want to cry every time I think of Earth Day in 1970, when we buried a car in high school. Who would have thought that thirty-six years later, people would be driving around in Hummers as a status symbol? Thomas Friedman says it all in his op-ed piece "In My Next Life" in the *New York Times*: "Then I at least want to be the owner of a Hummer—with American flag decals all over the back bumper, because Hummer owners are, on average, a little more patriotic than you and me. Yes, I want to drive the mother of all gas-guzzlers that gets so little mileage you have to drive from gas station to gas station. Yes, I want to drive my Hummer and never have to think that by consuming so much oil, I am making transfer payments to the worst Arab regimes that transfer money to Islamic charities that transfer money to madrassas that teach children intolerance, antipluralism and how to hate the infidels. And when one day one of these madrassa graduates goes off and joins the jihad in Falluja and kills my neighbor's son, who is in the U.S. Army Rangers, I want to drive to his funeral in my Hummer . . . stopping at two gas stations along the way."

Ethan: Those seem to be the dots you're trying to connect in your show . . .

Jon: Yes, the show is about connections, connecting the dots of causality. Some of this is created by the camera movement combined with the machines acting as editing machines, but also from the viewer's movements and cognitive understanding of the show. The belligerent insistence to start a war; antiwar protests; consumer culture, which keeps everyone dazed and confused; Shock and Awe; Bush watching Shock and Awe on TV; you watching Bush watching; someone watching you watching Bush . . . In New York, the final piece I made was *The Drowned World*, which refers to Hurricane Katrina, a tragedy that exposed the fact that we can't answer the needs of our own citizens, especially if they're poor and black, and that the government is more anxious to pay for civic construction in Baghdad, even if not much is built and especially if the money goes to Halliburton. A touchstone of this show was the film *Apocalypse Now* and Coppola's desire to play with and personalize the history of the Vietnam War, creating narratives that subjectively expressed the insanity of the situation. The difference, of course, was that *Apocalypse Now* was made after the war. That's why I was so struck by the television show *Over There*, the scripted docudrama about our troops in Iraq as it was occurring. There's no better example of the immediacy of the absorption of life into mediated experience. I believe we have come to a point where there is a separation from experience and the image; the image has become liberated from the experience, such that through the news, the Internet, and various other media, we encounter an overload of imagery, what Paul Virilio calls a visual crash.

General: I'm so sick of cameras everywhere I go—being filmed all the time. I suppose it could be a deterrent for perpetrators, or maybe we're gathering more visual information than we could possibly use. Even so, the military is at the forefront of the intelligent processing of all of this audio and visual information.

Ethan: Yeah, with AT&T's help and personal rights being sacrificed . . .

General: You know what they say, sonny boy, you have to break some eggs if you want to make an omelette. Jon, can you talk about the use of surveillance in the work?

Jon: The constituent character of surveillance—mediated cultural experiences augmented by anxiety—is present throughout all my work since 9/11. However, "The Palace at 4 A.M." and some of my newer pieces respond specifically to the Iraq war and the unlawful occupation of that country. Unlike "Global Village Idiot," which was composed of eight or nine discrete sculptures, "The Palace at 4 A.M." implicates the spectators, turns them into the surveilled subjects, and refers to our ambiguous role in these recent world events. The play of camera, image, and monitor is ubiquitous, and spectators are constantly disrupting sight lines, thereby becoming active viewers/participants in the installation.

Ethan: As much as the show addresses surveillance and the media, it also insists on an unmediated experience . . .

Jon: The show is meant to be experienced. There is heat generated by the television sets, the smell of rubber from the miles of electric and coaxial wire, the sounds of the snipers imitating gunfire. There is a wave of experiential stimuli, which leads the viewer through the installation. While "The Palace" conceptually foregrounds the role of and our relationship to the media, it is in many ways a tribute to the unmediated experience. The machines are a physical experience for the viewer to actually have, quite different from viewing videos in a dark room.

Ethan: And yet it comments brilliantly on the image production of our time.

Jon: One of the show's intentions is to oversaturate the viewer's visual stimuli and expose the world as a prop for the constant fabrication of images to feed our collective desires. Reality shows and photo-op wars are an unambiguous manifestation of this phenomenon and an example of the democratization of voyeurism. The exhibition is emblematic of our historical moment, where time and imagery are conflated, so that our relationship to experience becomes increasingly confused and distorted. This complexity is internalized by the viewer, who simultaneously becomes spectator, performer, voyeur, and exhibitionist. If we are to appreciate our infatuation with and proximity to surveillance, then, for me, the question becomes not *How can we destroy the camera?* but *How can we undermine the surveilled image and empower ourselves?*

Fig. 1

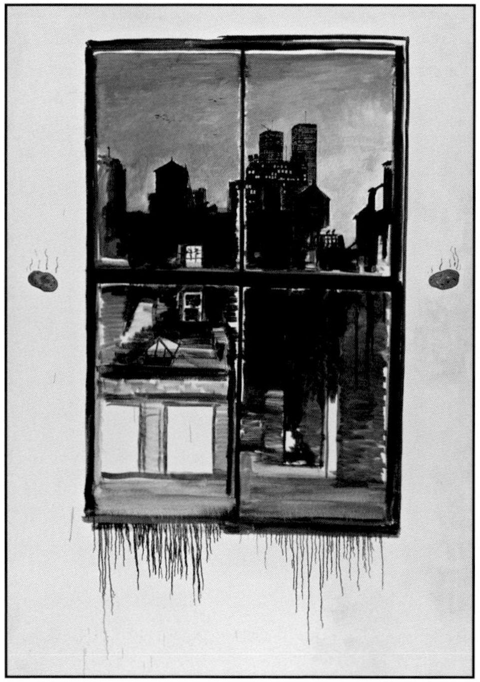

Fig. 2

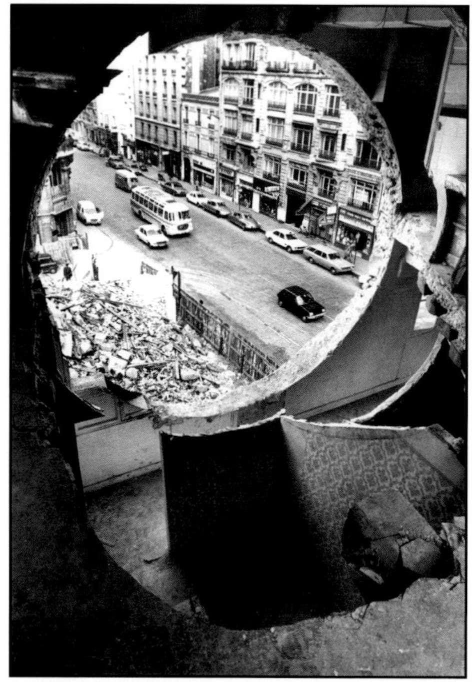

Fig. 3

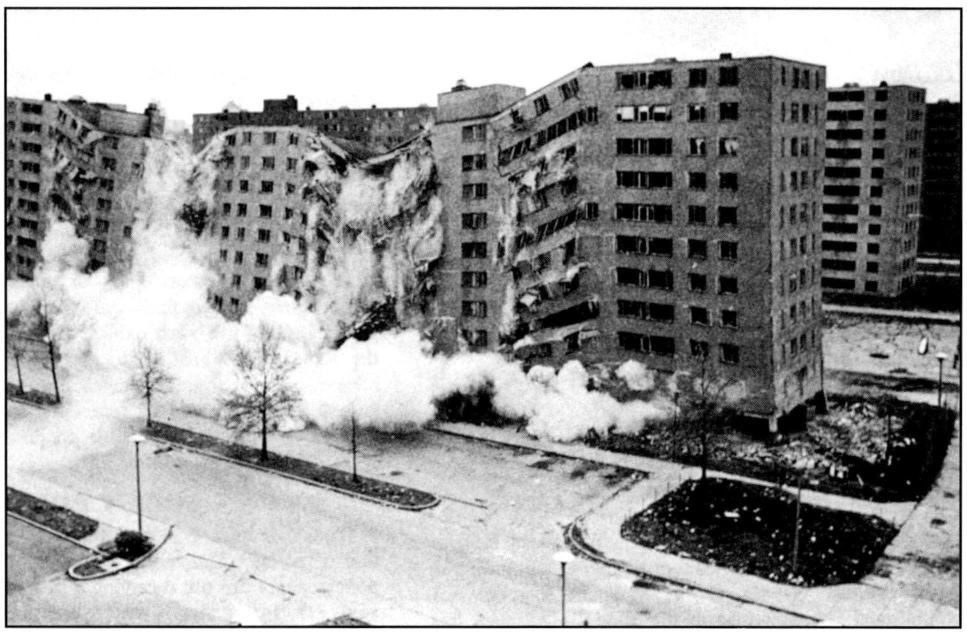

Fig. 4

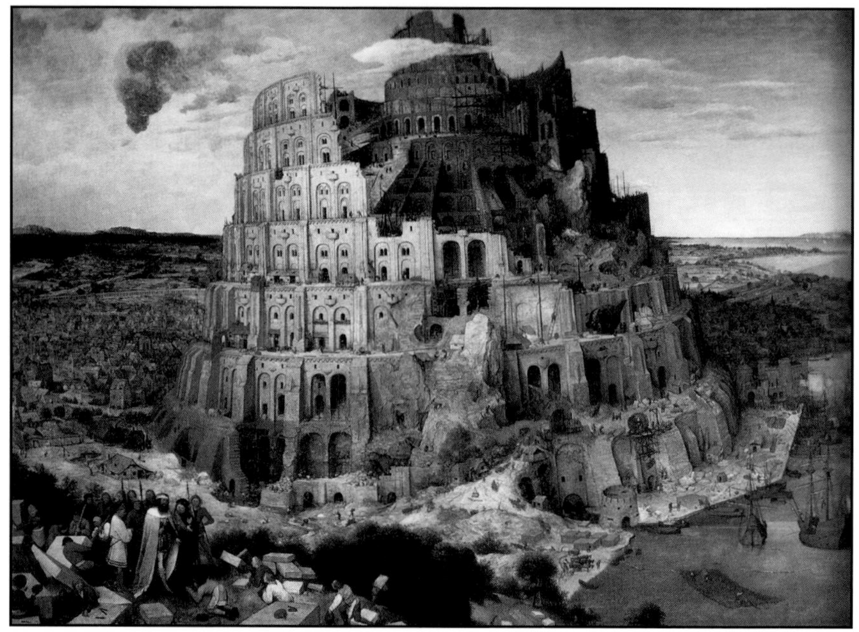

Harald Falckenberg

Journey Through Time, Back to the Origins of Identity

"Germany has declared war on Russia. Afternoon, swimming lesson."
(Franz Kafka, journal entry, August 2, 1914)

Jon Kessler's large-scale multimedia show "The Palace at 4 A.M." at the P.S.1 Contemporary Art Center in 2005 was a tremendous success. When the Düsseldorf gallerist Hans Mayer called me from New York to suggest I show the installation at my factory space in Hamburg-Harburg, I spontaneously agreed. So far, all I knew were details. According to Mayer's report, it was about Bush and a critical analysis of the Iraq war. Only as the installation was being set up did it become clear to Kessler and myself what we had let ourselves in for. The starting point for the work is the idea of capturing the terrorists' final view of the World Trade Center before they crashed into it. The footage of the planes hitting the Twin Towers runs through the installation from beginning to end.

As a work dealing with 9/11 and its consequences, its installation in Harburg is a return to its roots. Marienstrasse, just a few hundred meters from the exhibition venue, is where the attacks were planned by Mohamed Atta and his accomplices. When Kessler learned this, he immediately installed a surveillance camera pointing toward Marienstrasse that sent pictures of all movements in the area to several monitors spread around the installation. Further, he mounted an airplane cabin's molded-plastic window frame, an object he claimed had fallen into his lap during his flight to Hamburg, on a window of the exhibition venue.

This was my second encounter with 9/11 in art in just a few weeks. I had just purchased several works by Paul Thek from the period around 1970, when Thek was involved in actively resisting the construction of the WTC. In drawings, he placed gigantic black crosses between the Twin Towers as a symbol of their demise. Like many other New Yorkers, Thek did not understand why a long-established, organically multicultural neighborhood should be flattened to make way for a center of the international capital market. A painting from 1972 (Fig. 1) shows a view from his studio of the towers shortly before their completion. The picture is surrounded by hot potatoes—a play on words referring both to the children's game where those left holding the potatoes when the music stops burn their hands and to the rotten stocks that continue to change hands at a profit until the market falters and their true value comes to light. In the works of both Thek and Kessler, as in the classic Hitchcock movie, the outside world is observed and monitored via a *Rear Window*, as an expression of voyeuristic and paranoid obsessions that lead to persecution mania.

And another connection. In April 2007, the Centre Pompidou celebrated its thirtieth birthday with an exhibition entitled "Air de Paris." The show hinged on Gordon Matta-Clark's 16mm film *Conical Intersect*, 1974, which documents a "cutting" action by the artist, who, with the Situationists, was a bitter opponent of the Beaubourg project. Matta-Clark cut through the outer facades and interiors of two condemned houses in the immediate vicinity of the new building in an axis trained on the walls of the future museum. (Fig. 2) Although Matta-Clark could only gain authorization to carry out his cuttings on buildings that had already been condemned, he said he would "just as soon deal with something that's brand new, crisp, and not ready for the ax."[1]

The link between these two events—apparently far removed both in time and in cause—and the current situation is striking. Coincidence? One should be careful with the use of terms like *coincidence*, *situation*, and *event*. More than a few people consider coincidence to be the rule, order the exception.[2] This view becomes more plausible if order is considered to include only deliberate, logically comprehensible constructs. According to the latest findings of cognitive psychology and neuroscience, cerebral space is subject to constant change due to the reorganization and rewriting of memory traces, with coded, repressed, and delayed subjective and collective meanings being moved around, the vast majority of which—scientists speak of up to 90 percent of all perceptive and cognitive processes—takes place below the threshold of consciousness.[3] In the model put forward by Gilles Deleuze and Félix Guattari, art, science, and philosophy, as the three main forms of organized thought, constitute planes that cut through chaos to prevent the human mind from being destroyed by it.[4] It is always a matter of reaching the shore safely, combined with hopes that risk being submerged by the next tidal wave. It is the distinguished task of the artist to run through man's illusions and delusions in a finite context, all the more effectively to expose them, to break the illusions down, and to be free for a creative praxis beyond what is merely contemporary and fashionable.

Relations between architects and artists have always been fraught. On the subject of museum architecture and art alone, many books have been written. Architecture realizes the client's wishes for significance and greatness; art struggles for independence and autonomy. Although barely acceptable in human and social terms when the victims are taken into account, it was thus only logical that artists like Karlheinz Stockhausen and Damien Hirst celebrated the demolition of the Twin Towers as a torch lit against capitalism. Writing on Gordon Matta-Clark, Dan Graham put it in a nutshell in the mid-1980s: "The architect builds, the artist destroys."[5]

It would have been interesting to hear the opinion of Minoru Yamasaki (1912–1986), the Japanese-American architect of the WTC. He was spared having to comment on the demise of his greatest work, which is probably just as well. Almost thirty years previously, the demolition of the Pruitt-Igoe housing project in St. Louis, designed by Yamasaki in 1950 and dynamited by public order in July 1972, made deconstructive history. (Fig. 3) According to the official version, deteriorating social conditions were the reason for the demolition. But what was actually at stake was a radical turning away from ideological notions that had seen housing projects as the new home for a modern classless society. In their works heralding postmodernism, Charles Jencks and Tom Wolfe cast Pruitt-Igoe as a symbol for the end of modernism. In 1977, Jencks gave the opening section of his *Language of Post-Modern Architecture* the concise title "The Death of Modern Architecture." In 1981, Wolfe settled his own scores in *From Bauhaus to Our House*. And Pruitt-Igoe made an impact in Germany, too: The name of Berlin experimental band Einstürzende Neubauten ("collapsing new buildings"), founded in 1980, was inspired by the housing project's topoclastic removal.

Jon Kessler's oeuvre stands in the tradition of postmodernism. A descendant of an established Jewish family from New York, he was expected to take over his father's clothing store. But Kessler wanted to be either an artist or a musician, and it turned out he became both. It was his intention to break through the mimetic adaptation of Jewish tradition to American culture. The "Broken Buildings," a little-known series of works made between 1989 and 1991, is related to architecture. Kessler dismisses the idea of deriving identity from shells, forms, and functional frameworks. He was more interested in the skeleton and the mechanisms behind the facades.

In this light, his "Broken Buildings" series is less different from the kinetic sculptures he has developed as his main work since 1983 than one might think at first sight. In these strange objects, Kessler works with components of multicultural origin that—commercially exploited as kitsch, knickknacks, and trash—are now nothing more than reflexes of long-past historical contexts. The cultural melting pot of New York is the ideal place for such metamorphoses. Kessler rejects the American dream of representation, power, and wealth. Unlike Jeff Koons, he does not understand kitsch as the polished hunting trophies of an ostentatious middle class. He resists the cold ideas of Pop and Appropriation art and rejects Nouveau Réalisme's relentless free play with objets trouvés. His handling of objects is marked by a respect for the historical models and by the romantic idea of saving them from the trash can of desire. Recombined and reanimated, they make up random and strangely unfamiliar documents of an everyday life devoted to the past. "I'm not a chess player," Kessler says, "I don't make art that way, and I never know where a body of work will go until I get there."[6]

The aura of anti-authority and settling scores with the world of the fathers is a central theme. Louise Bourgeois has said: "You have to forget your past every day. And accept it. And if you can't accept it, you have to make sculptures."[7] She understands sculpture as unbridled pleasure in deconstruction and the symbolic destruction of the father—the cause of a constantly rekindled childhood trauma. Kessler's work of the mid-1980s can hardly be seen as anything other than an oedipal serial violation. The art world with its theoretical discourses remains alien to him. He sees himself as a loner: "I've always held a position outside of the line of fire. It's a good place to be: it's a front row center seat."[8] Kessler set himself apart in a fictional world of esotericism and contemplation far from the fathers with their commercial entanglements. The retrospective show "Jon Kessler's Asia," at the Kestner-Gesellschaft in Hanover and the Neue Galerie am Landesmuseum Joanneum in Graz, brought this phase of Kessler's emancipation to a close in 1994.

For many artists, the early 1990s were a nightmare. Having long been pampered and beguiled during the art boom of the 1980s, once the market collapsed, they were dropped by their gallerists overnight. They formed self-help groups and struggled to survive as critics, curators, and writers. Many became teachers at art academies. This was also the right path for the outsider Jon Kessler. Since 1994, he has been teaching at Columbia University's School of the Arts, from 2000 through 2005 as head of the Visual Arts Division. In this sheltered academic environment, inspired by his work with the students, he was able to develop his art "outside of the line of fire." Increasingly, he focused on the media-theoretical aspects of his work.

With "The Palace at 4 A.M.," Kessler made a breakthrough. The obsessive, ego-psychologizing focus on the artist's own identity is left behind in favor of a freer awareness in which systems of observation and surveillance overlap right through to the recipient, who watches the observer observing. This step does justice to the fact that the observer himself is part of the reality and the system that he observes—a blend of objective and subjective reality. "The Palace at 4 A.M." is an eminently political work. It is, not least from Kessler's personal point of view, proof of a development that has become known in media theory as the *cognitive turn*. According to this model, perception passes not so much through the eye as a supposedly neutral and objective organ[9] as through cognition, controlled by the brain as the central organ of cognitive processes. In his treatise "Perception in the Technological Age,"[10] media theorist Peter Weibel sees the cognitive turn as being characterized by two fundamental experiences: first, that technological, machine-assisted perception enables the control of virtual bodies, movements, and spaces via ever more complex trompe l'oeil technology; and second, that the world becomes dependent on the observer. With this paradigm shift from passive registering to active shaping, the world perceived by the observer constitutes itself as a system of free construction.

"The Palace at 4 A.M." takes Kessler, as a creator of the Newest, on a journey through time to the Oldest, to Mesopotamia, the anthropological and topographic origin of our linguistic and symbolic culture and the location of the Tower of Babel, one of the earliest monuments to human hubris. (Fig. 4) There remains the question of Kessler's Jewish identity. I imagine his answer might be similar to that noted by Franz Kafka in his journal on January 8, 1914: "What have I in common with Jews? I have hardly anything in common with myself and should stand very quietly in a corner, content that I can breathe."

1. Interview with Liza Bear, "Gordon Matta-Clark: Splitting the Humphrey Street Building," in: *avalanche*, December 1974, pp. 34–37, reprinted in *Gordon Matta-Clark*, London/New York 2003, pp. 163–169.
2. For more on this, see Henning Ritter, *Frankfurter Allgemeine Zeitung*, April 7, 2007, especially with regard to Sigmund Freud's treatise "The Uncanny" (1919). And is it another coincidence? A couple of days after I finished this text on June 8, 2007, the American philosopher Richard Rorty, the master-thinker of coincidence, died. The ideas of his central work, *Contingency, Irony, and Solidarity* (1989), are confirmed by the latest findings of cognitive psychology and neurology, cf. Harald Falckenberg on the 150th anniversary of Sigmund Freud's birth, in: *Aus dem Maschinenraum der Kunst*, Hamburg 2007, p. 60ff.
3. Harald Falckenberg, op. cit., footnote 2.
4. Gilles Deleuze/Félix Guattari, *What Is Philosophy?*, Columbia University Press 1994, pp. 208, 210–11, 216–17.
5. Dan Graham, "Gordon Matta-Clark," in: Sabine Breitwieser, ed., *White Cube/Black Box*, exhibition catalogue, Generali Foundation, Vienna 1996, pp. 215–237.
6. Jon Kessler interviewed by Lynne Cooke, in: *Jon Kessler's Asia*, exhibition catalogue, Kestner-Gesellschaft/Neue Galerie am Landesmuseum Joanneum, Hanover/Graz 1994, p. 54.
7. Quoted from Claudia Steinberg writing on Louise Bourgeois, *Kunstzeitung*, 12/2006, p. 18.
8. Jon Kessler, op. cit., footnote 6.
9. In this sense, Brian O'Doherty, *Inside the White Cube: The Ideology of the Gallery Space* (1976), expanded edition, University of California Press 2000, p. 16ff.
10. (2000), in: *Gamma und Amplitude*, Hamburg 2004, pp. 320–344.

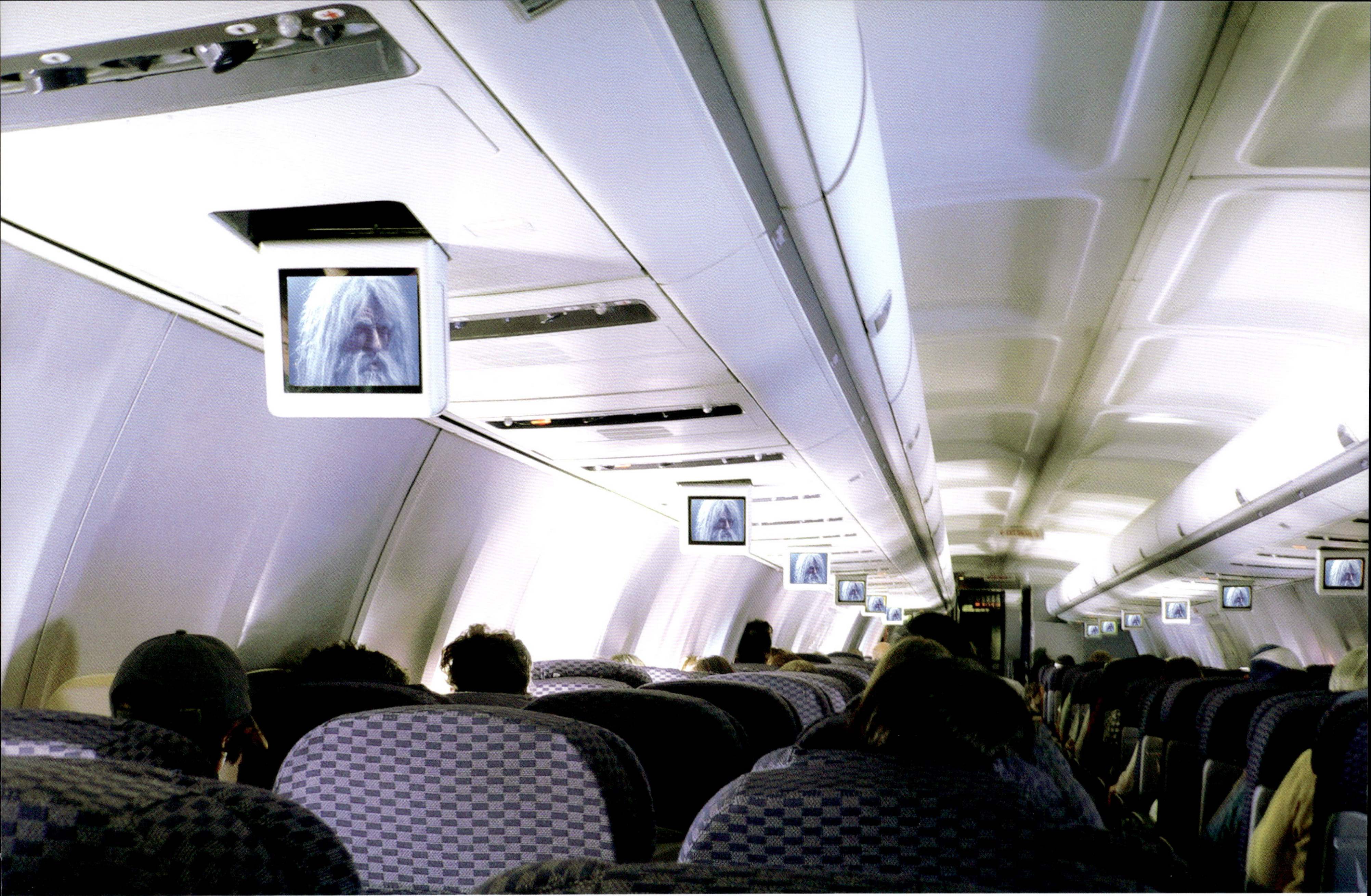

Hamburg

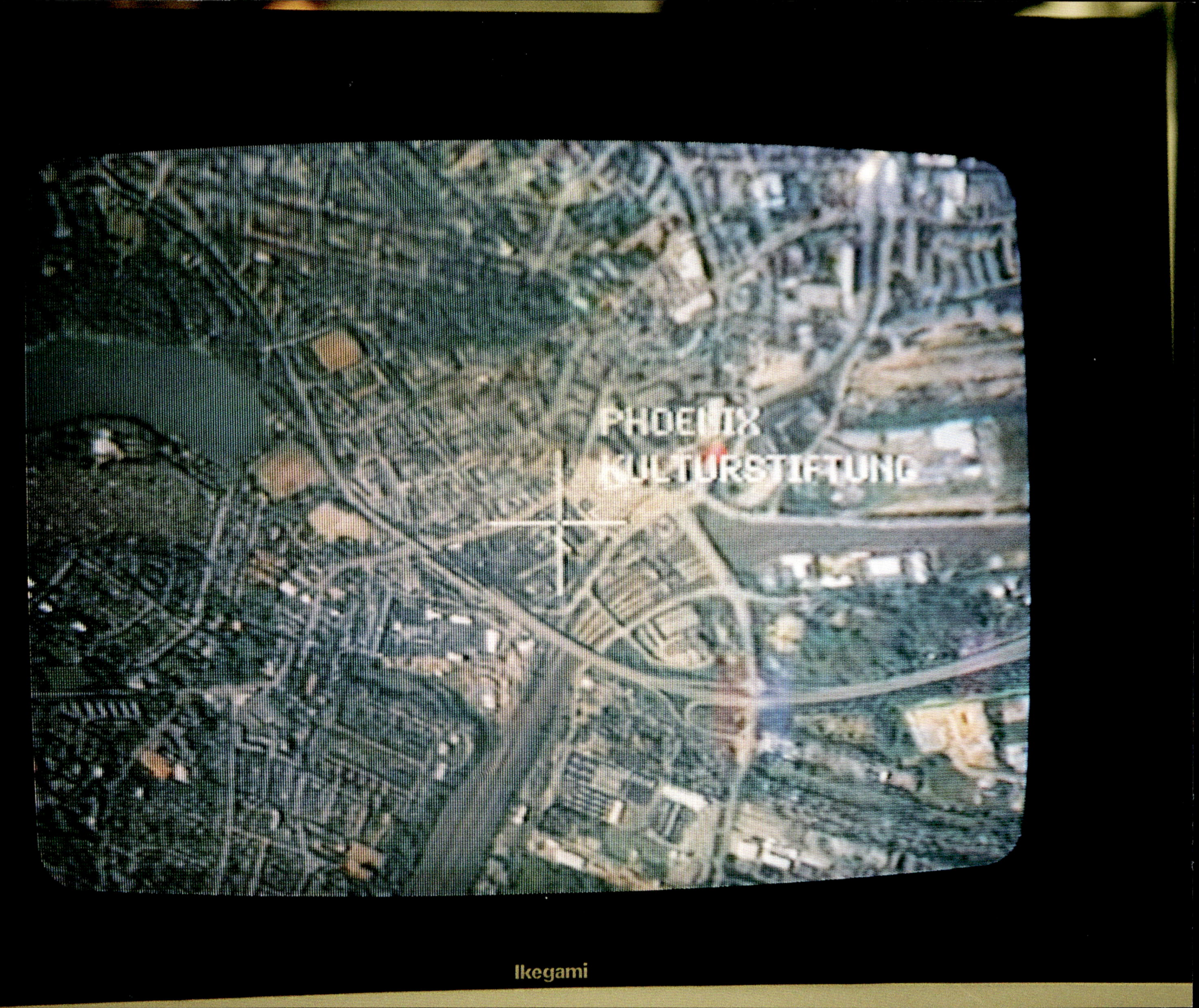
PHOENIX
KULTURSTIFTUNG
Ikegami

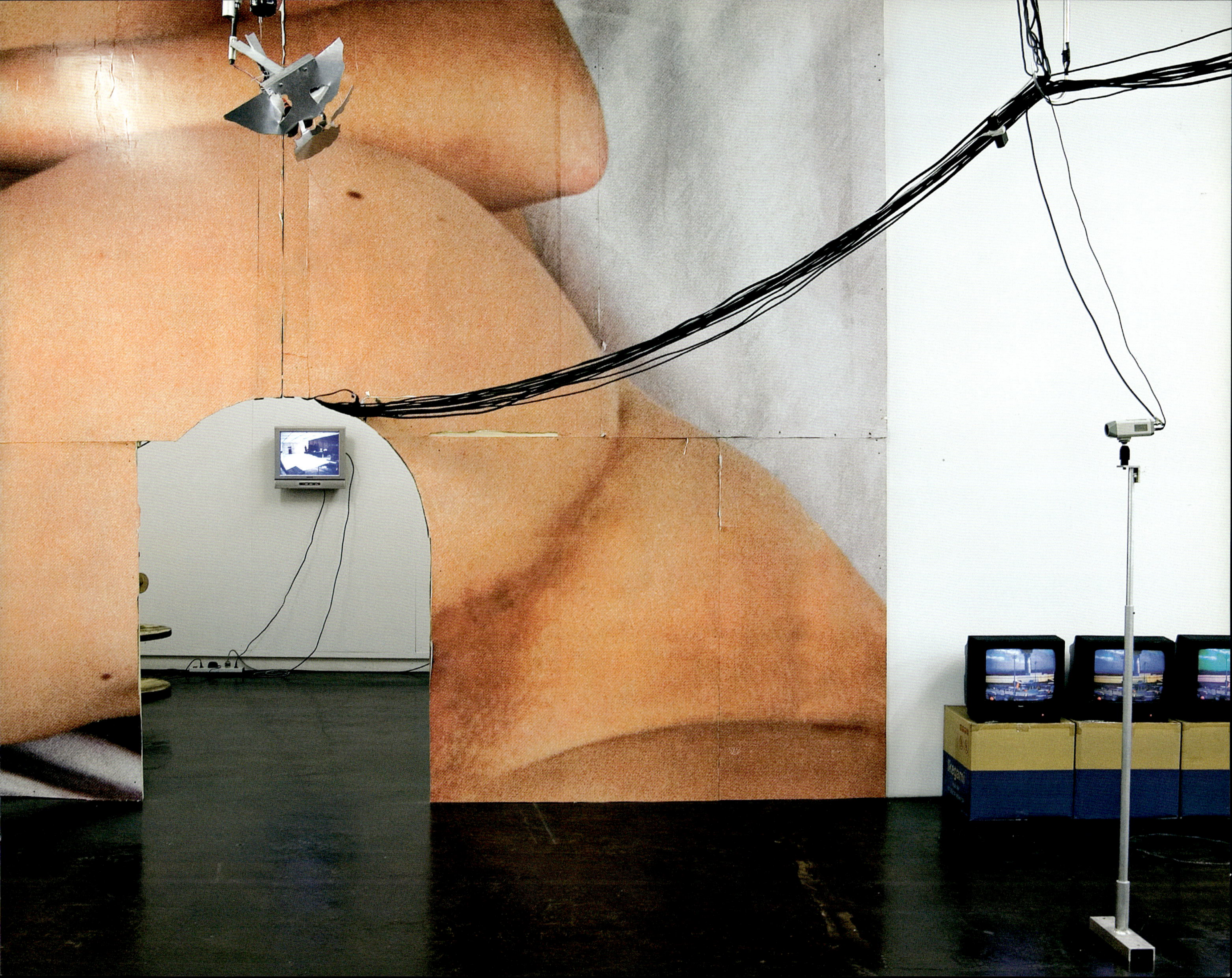

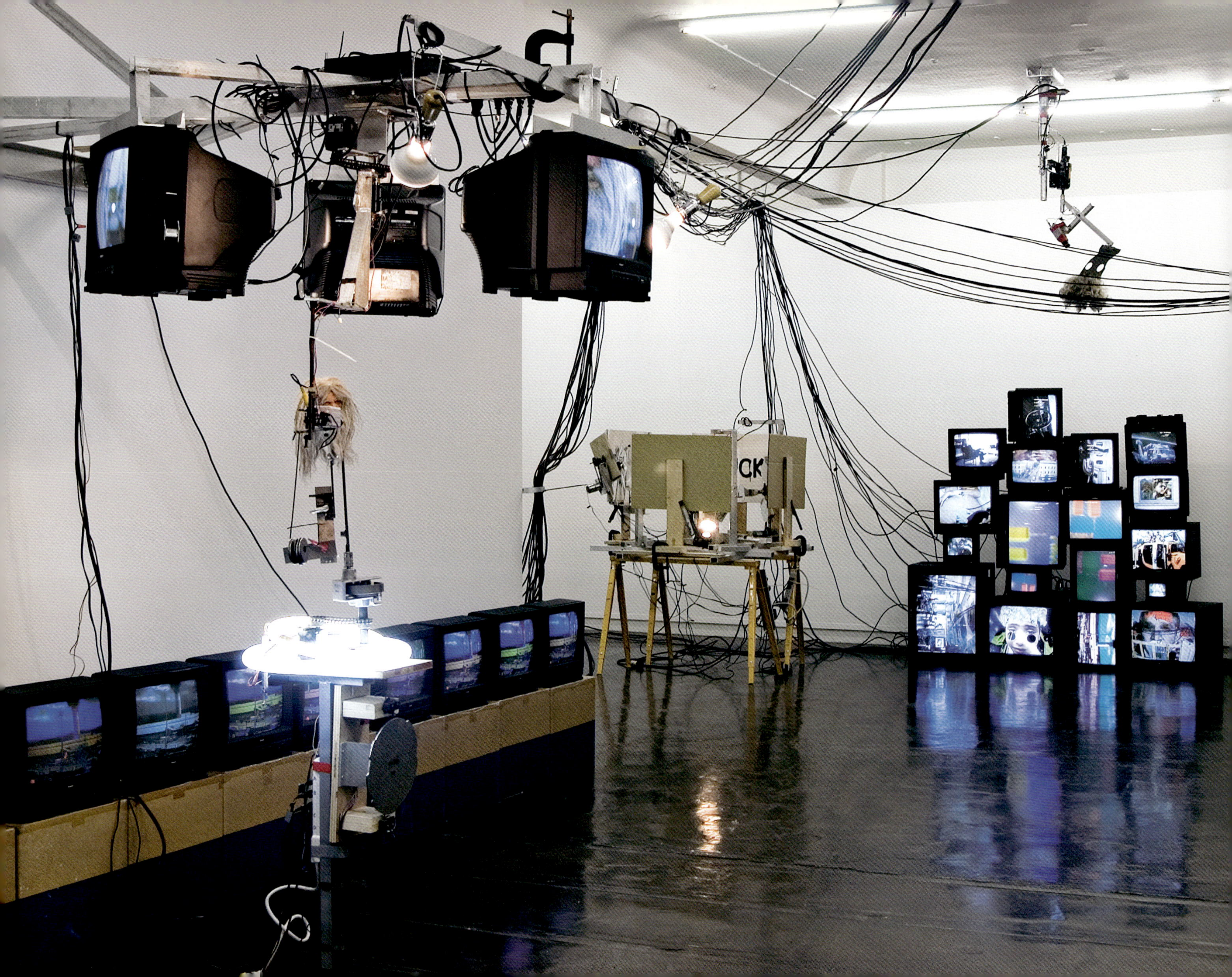

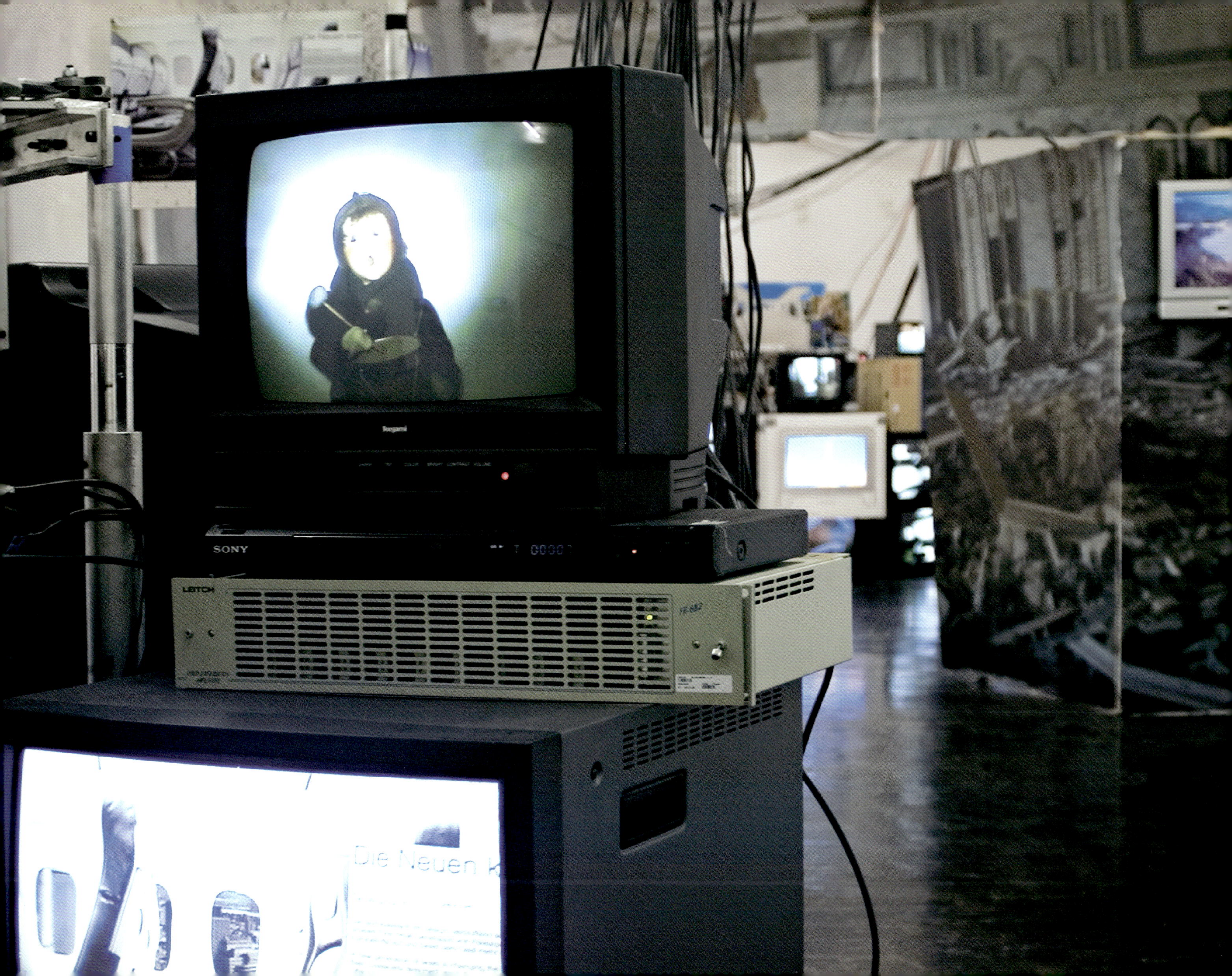

SONY
LEITCH
FR-682
Die Neuen

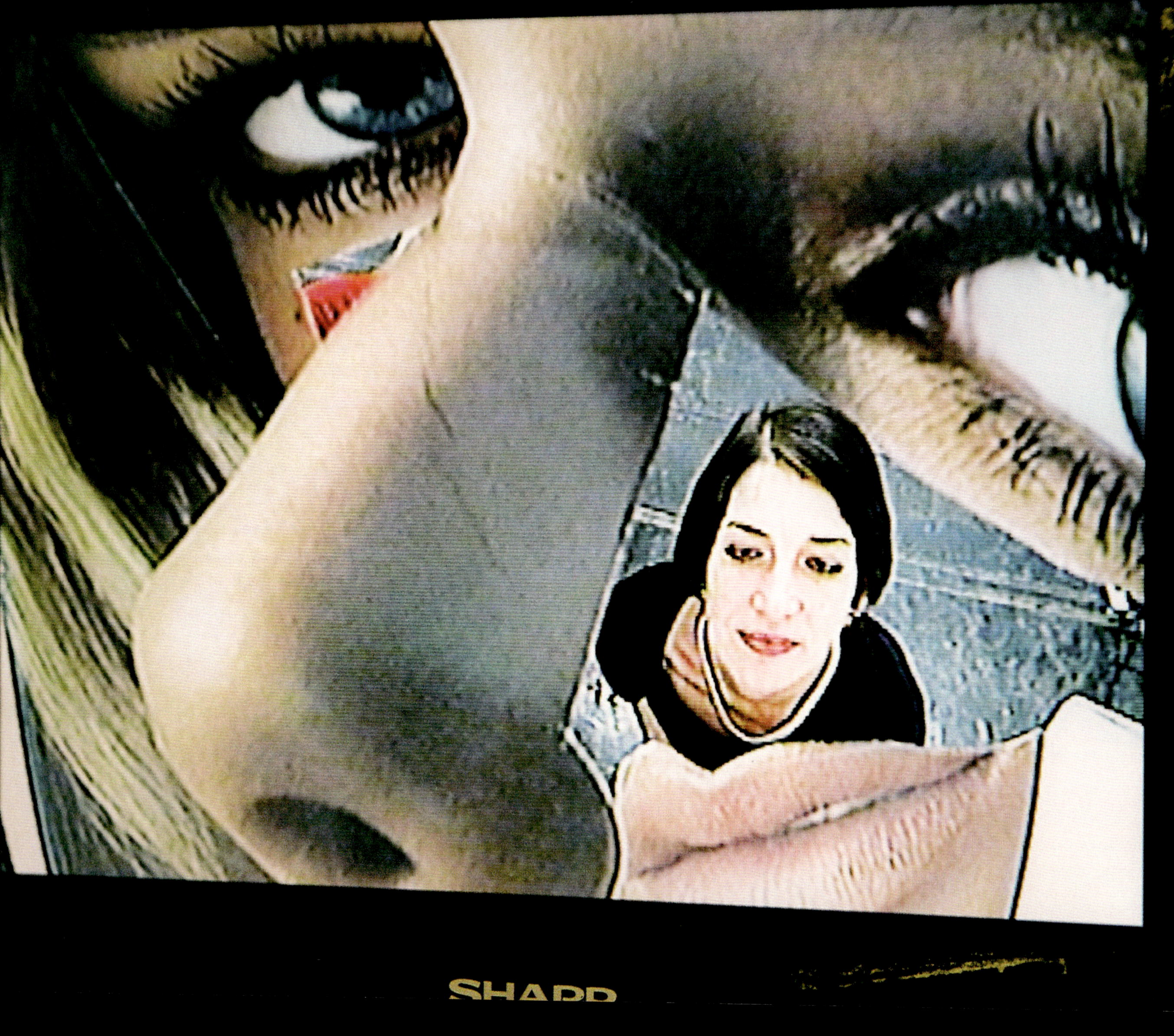

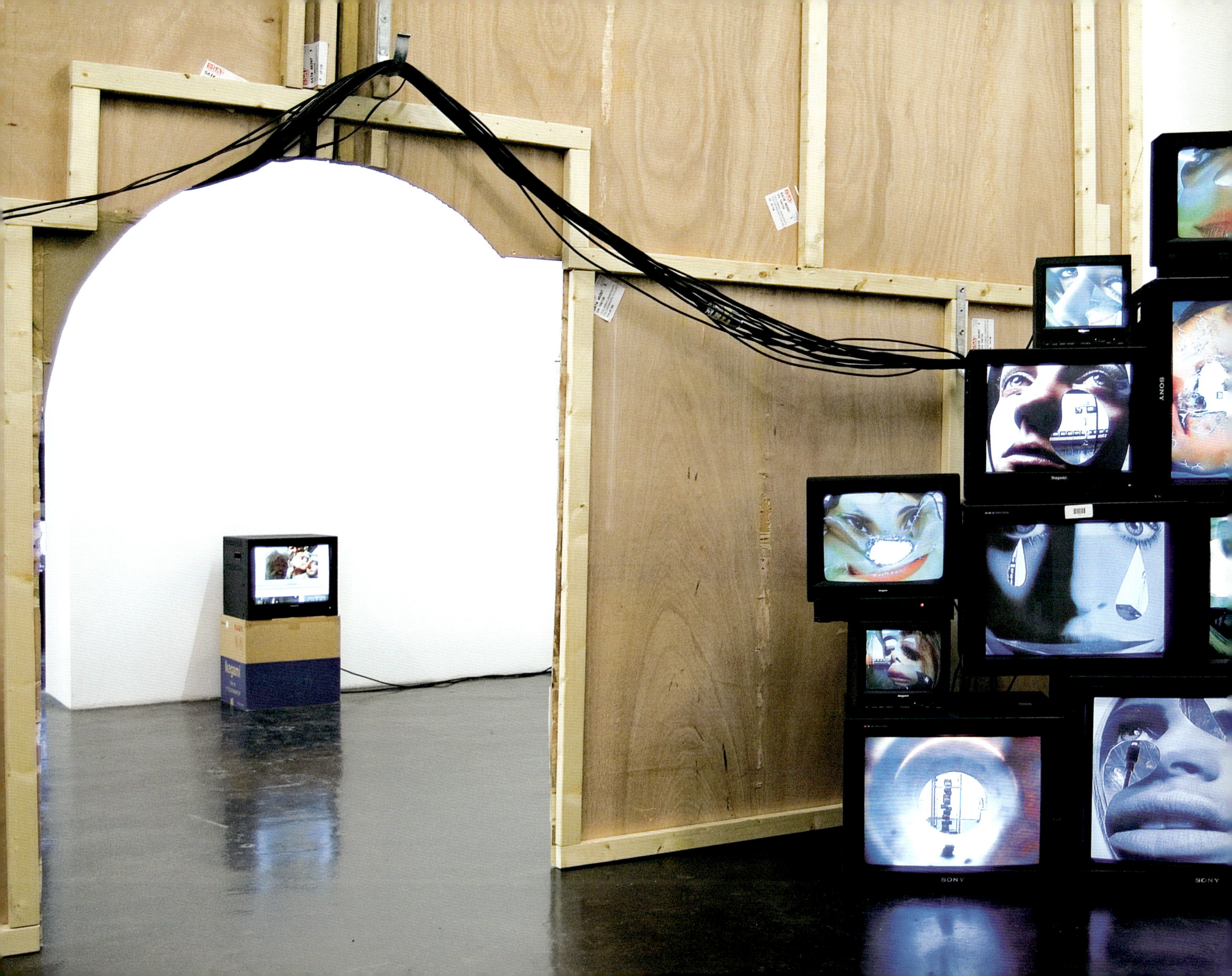
SONY
SONY
SONY

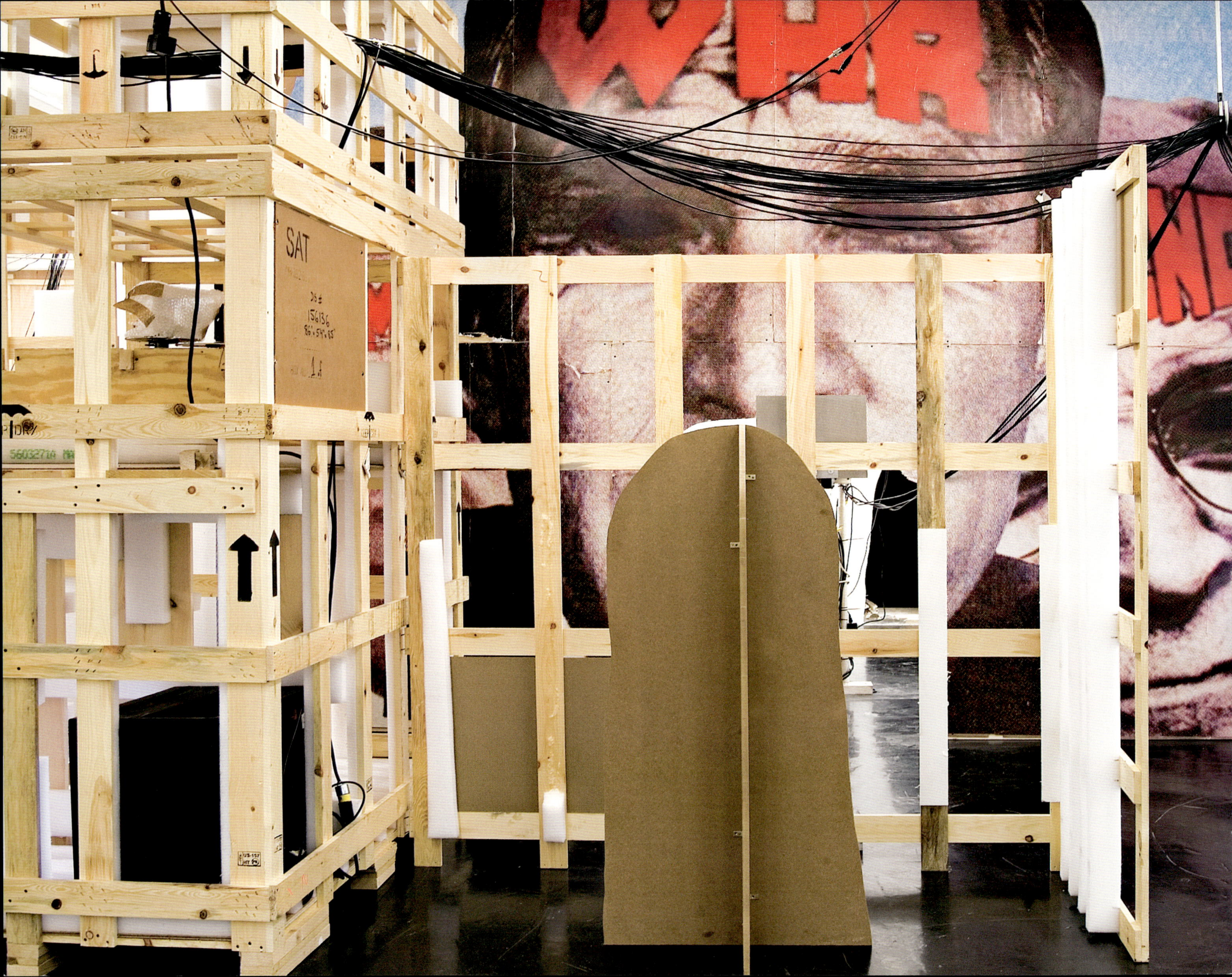
WHR
SAT

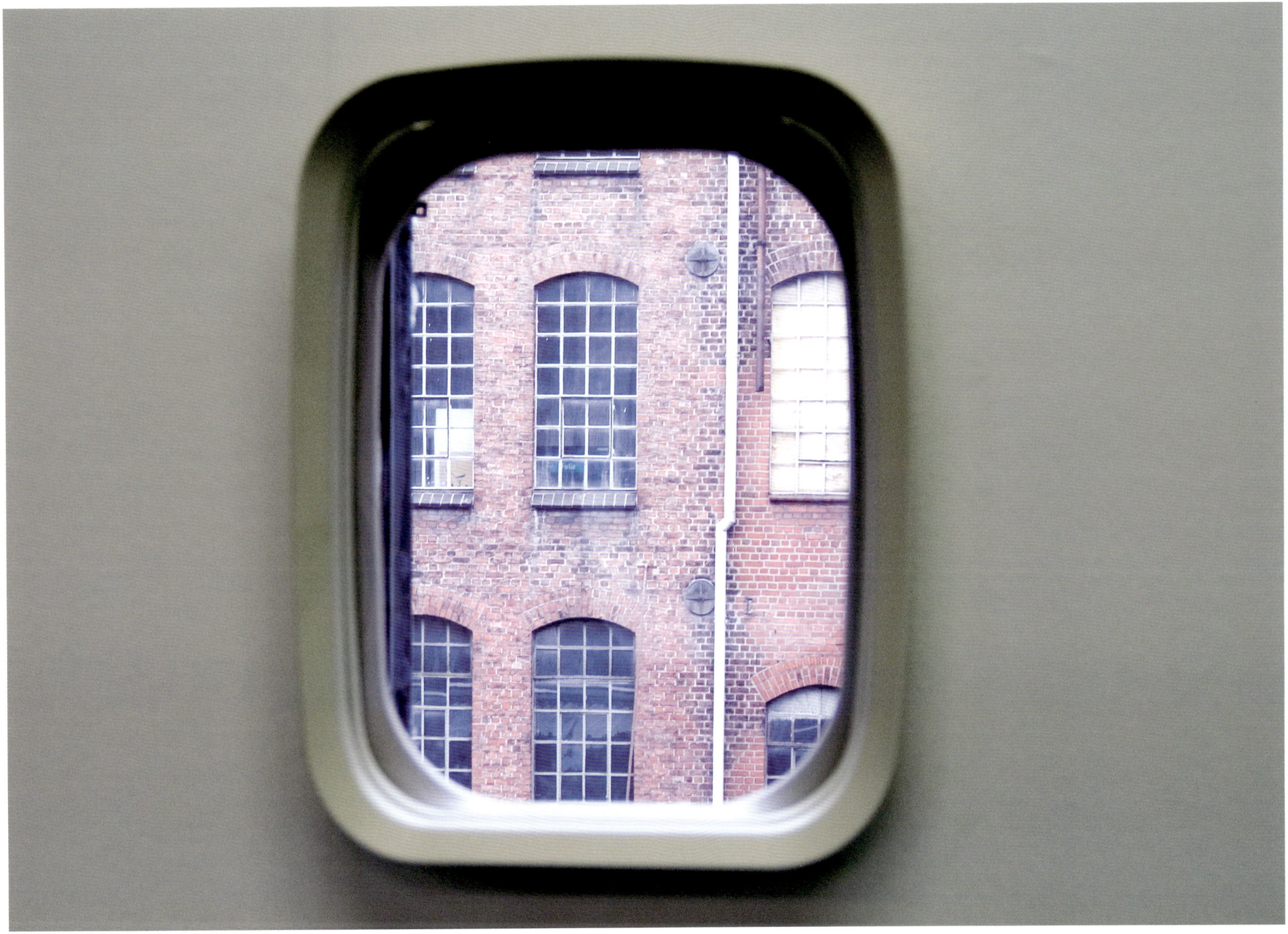

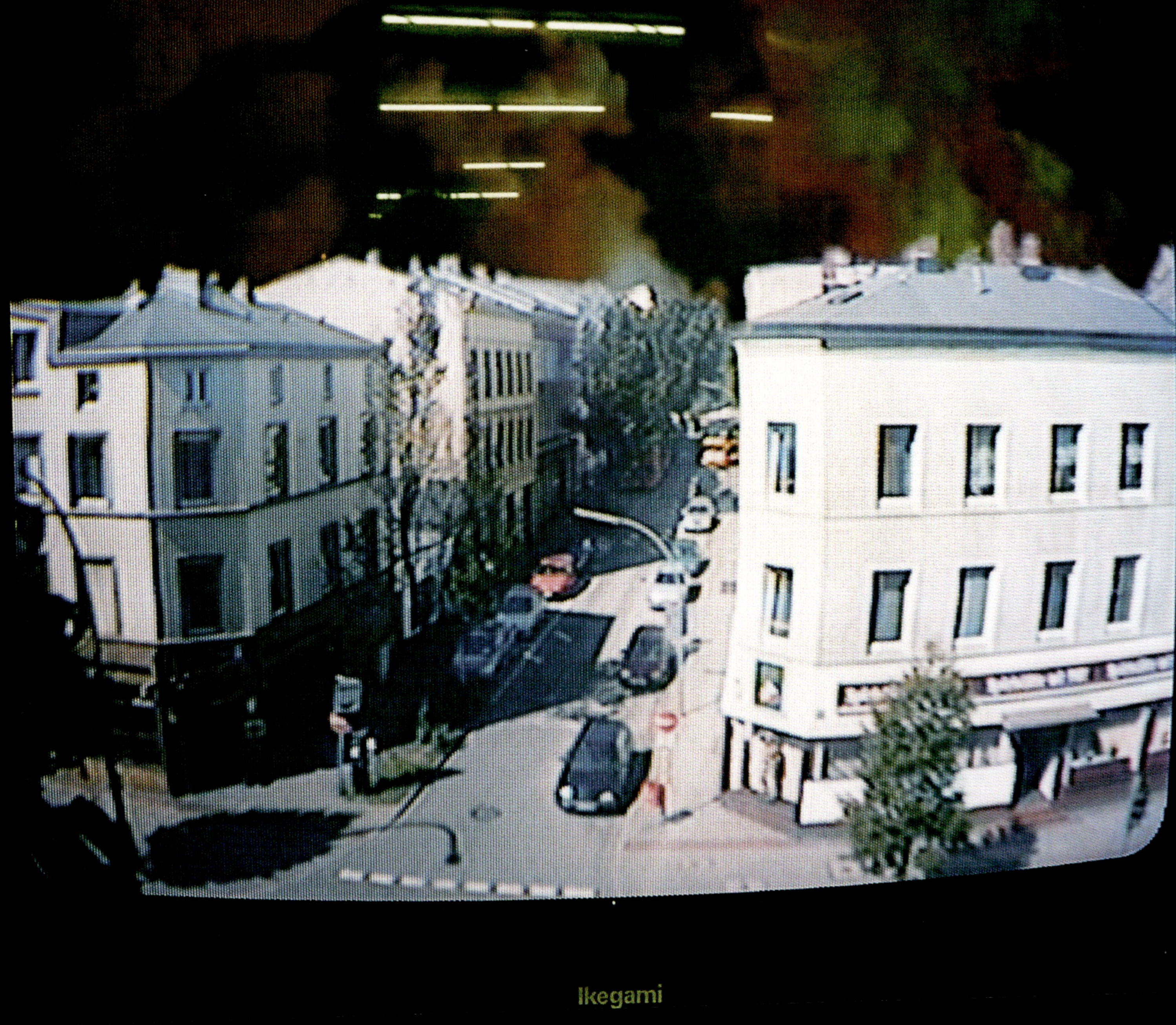
Ikegami

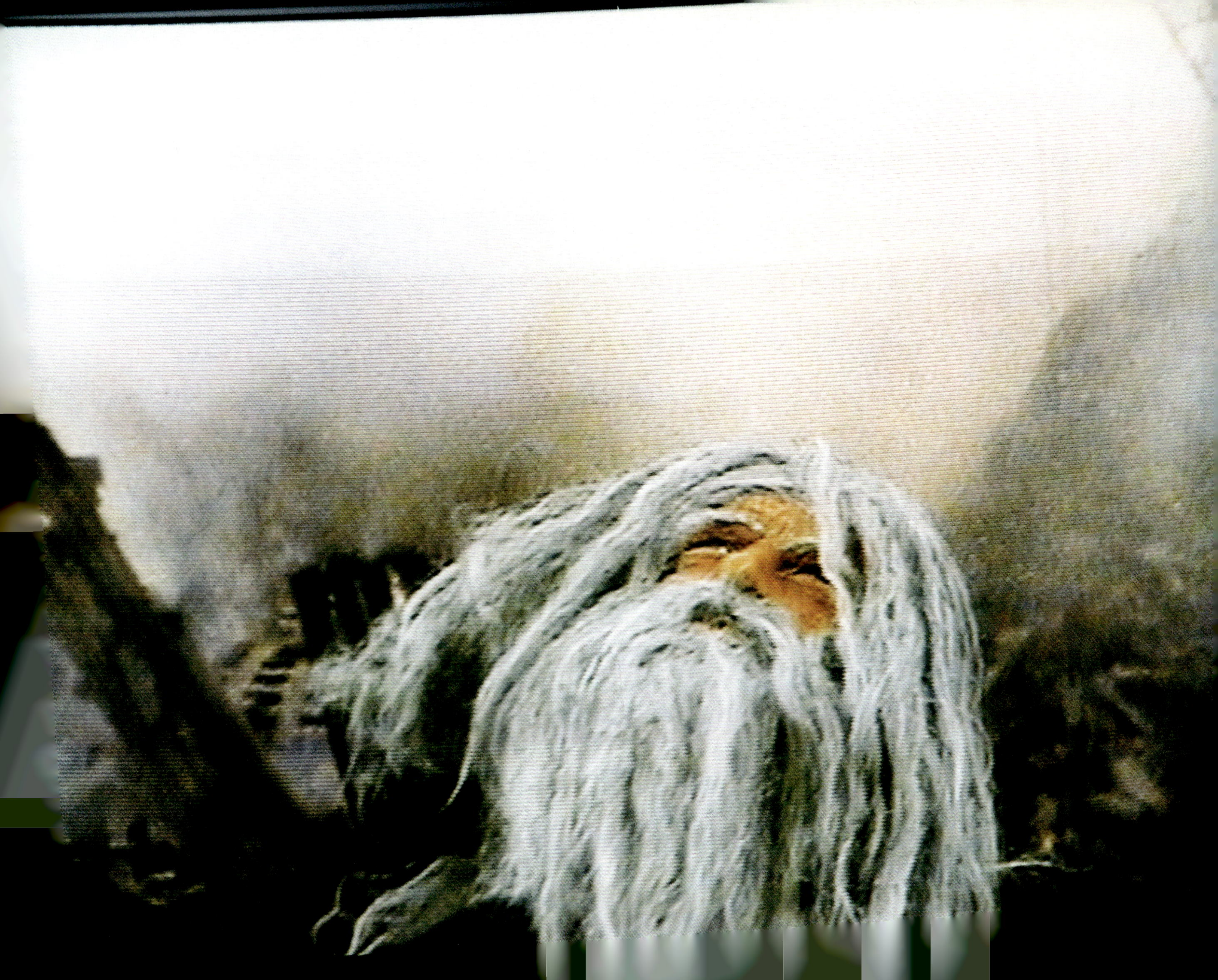

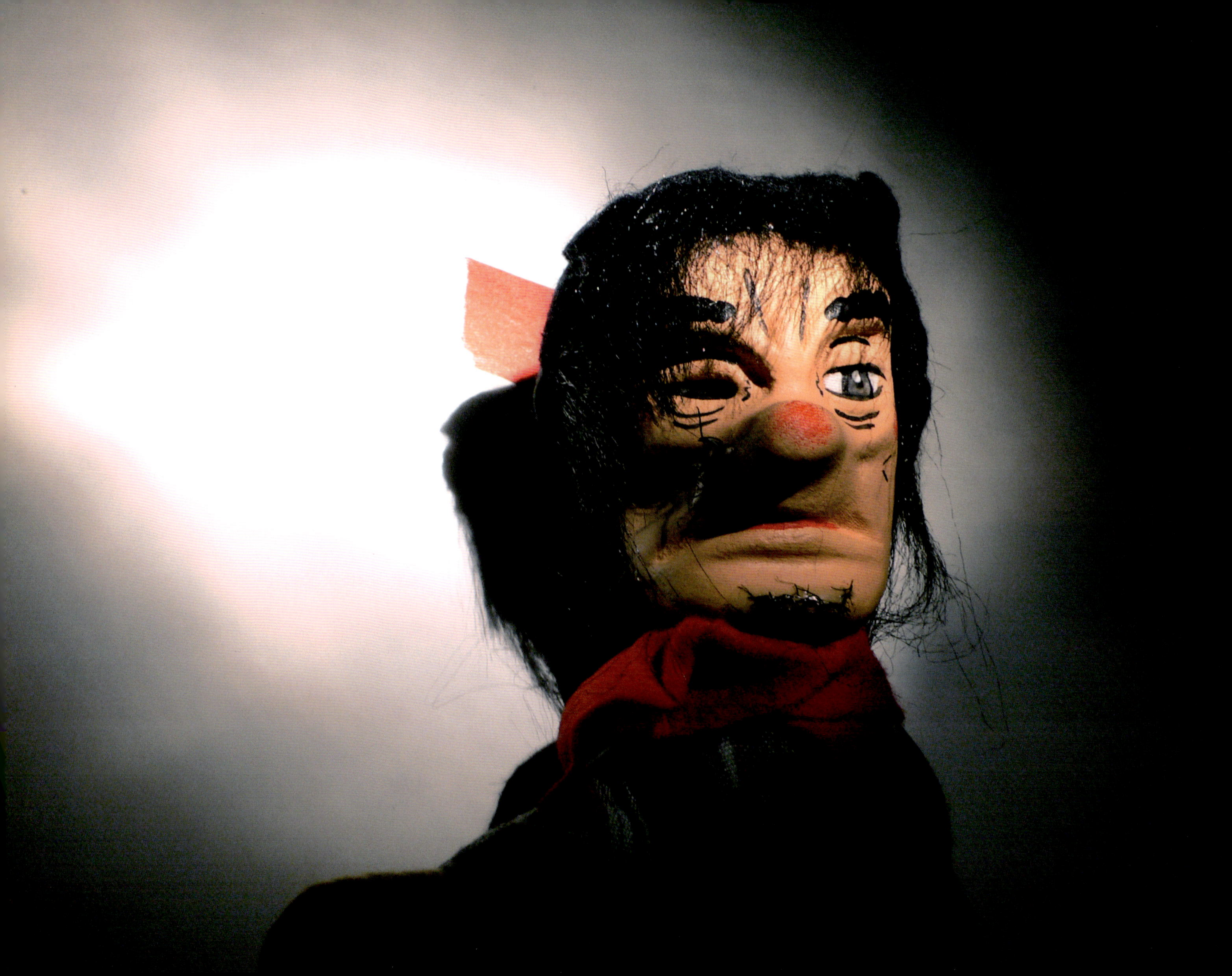

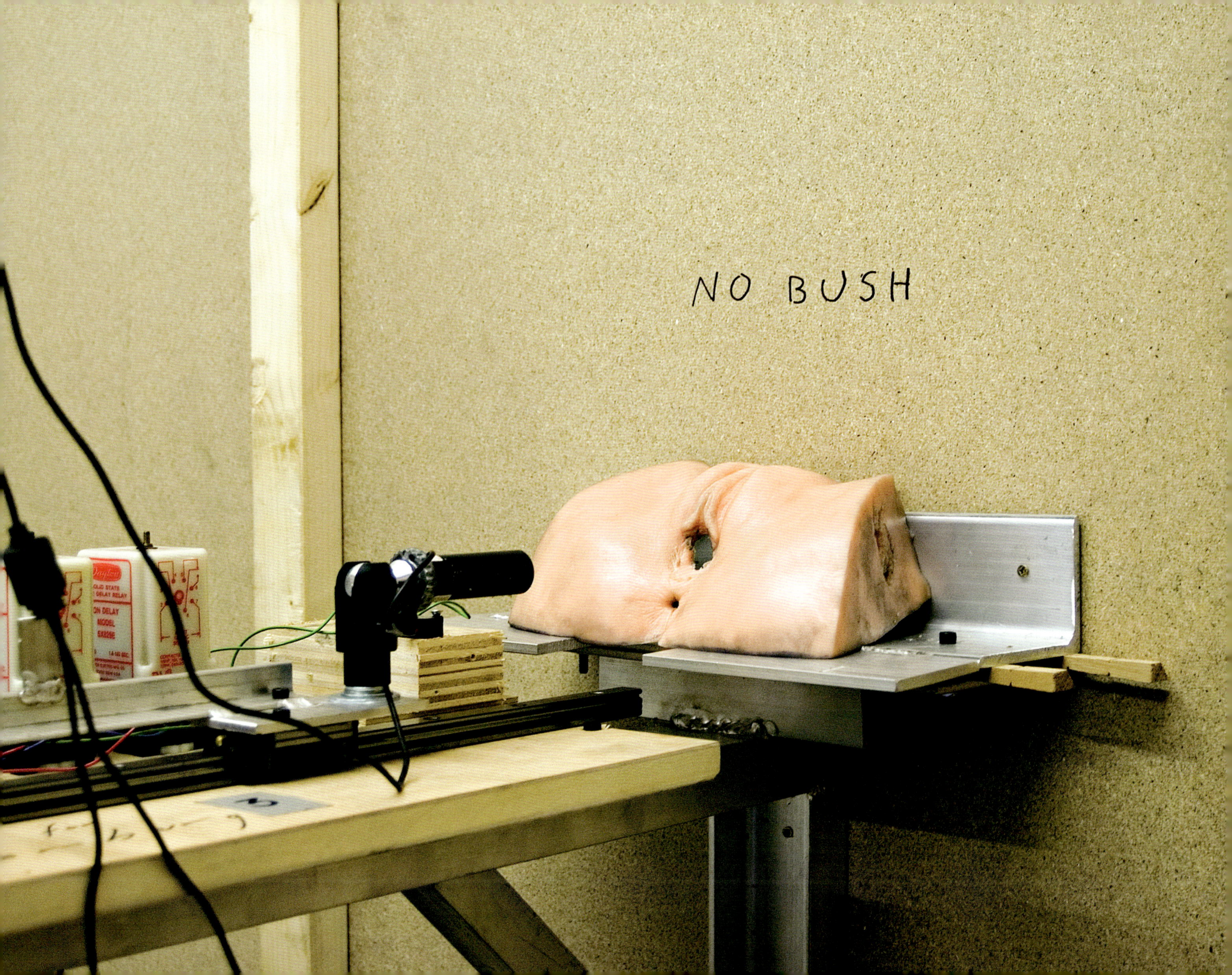
NO BUSH

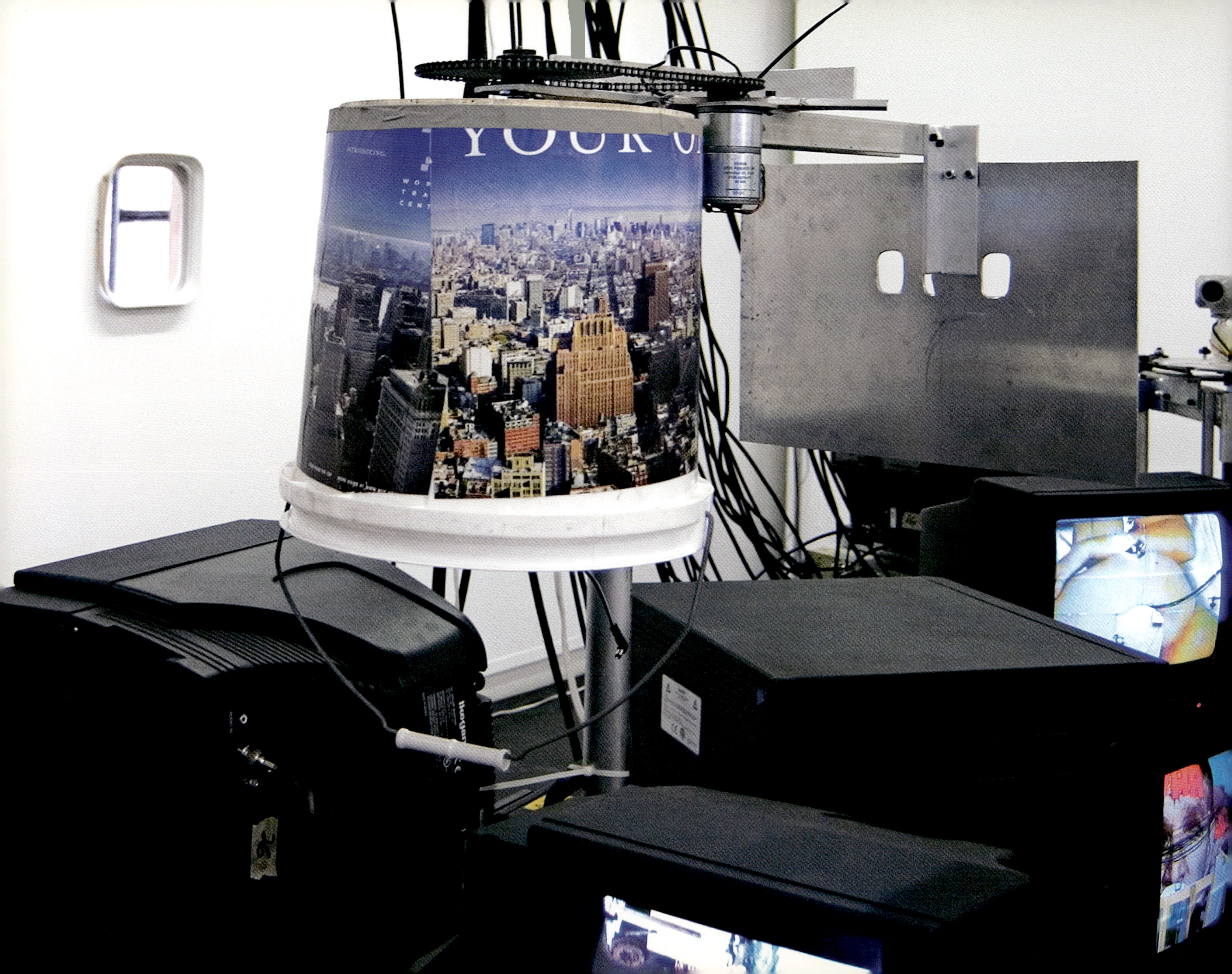
YOUR

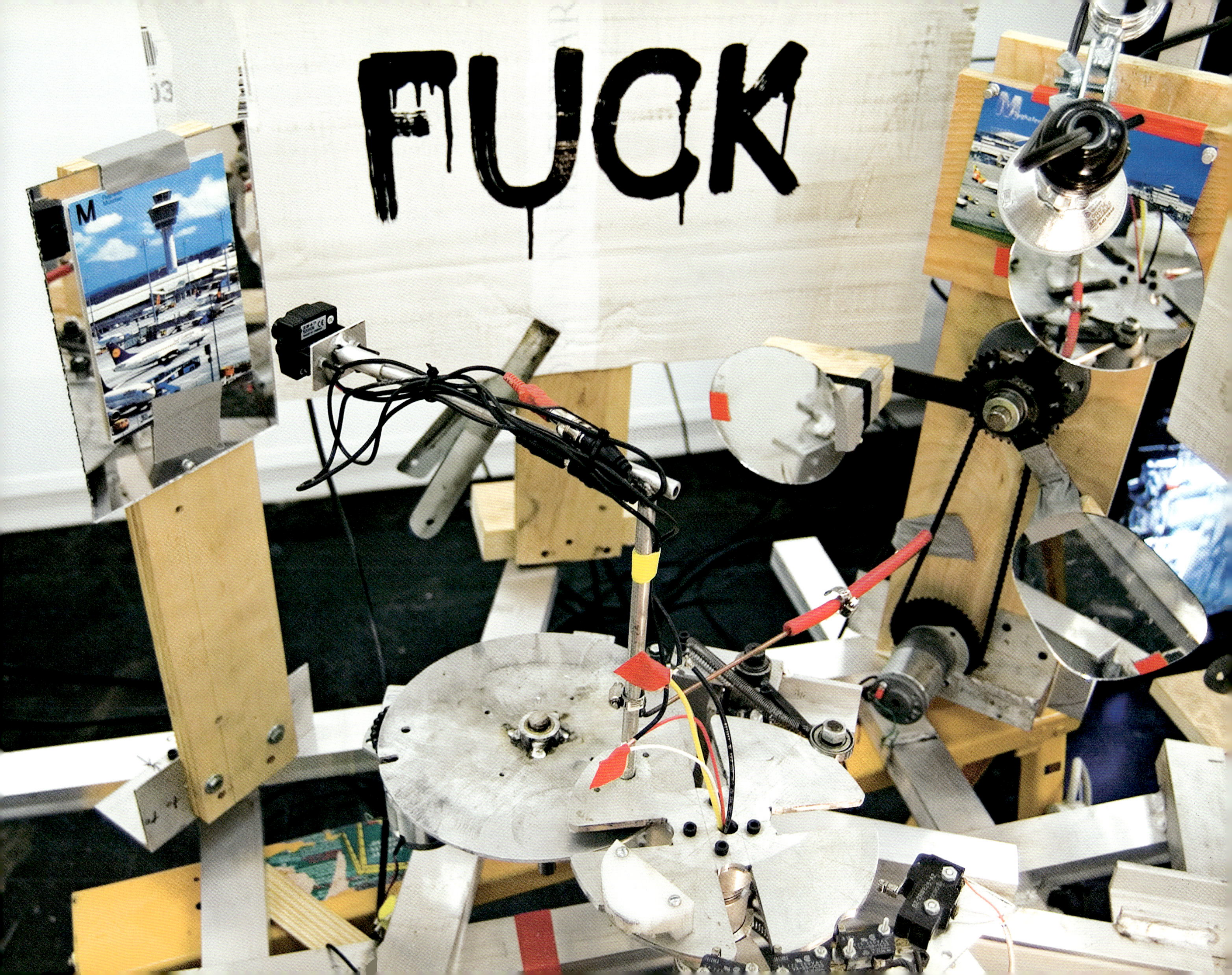
FUCK

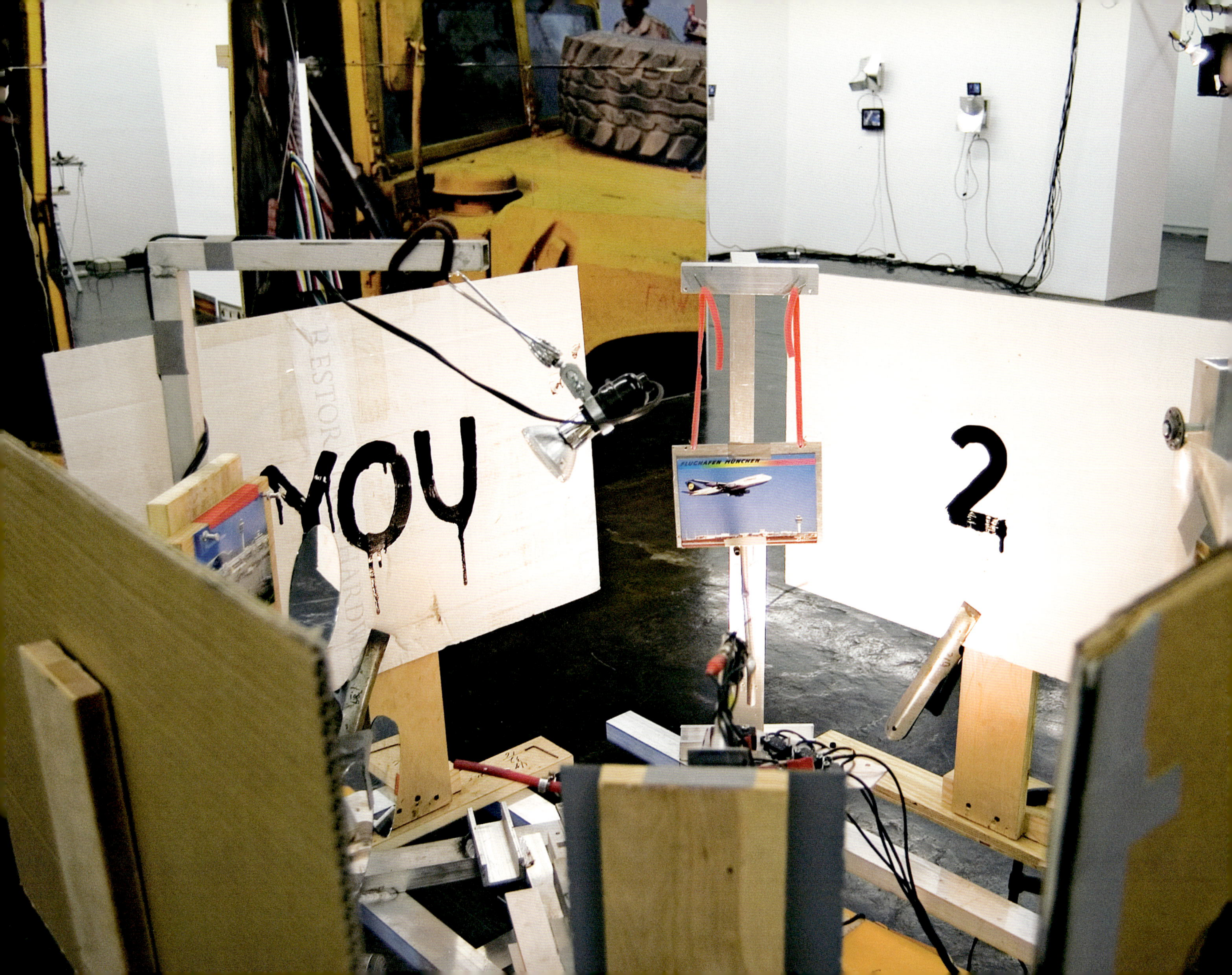
YOU
2
FLUGHAFEN MÜNCHEN

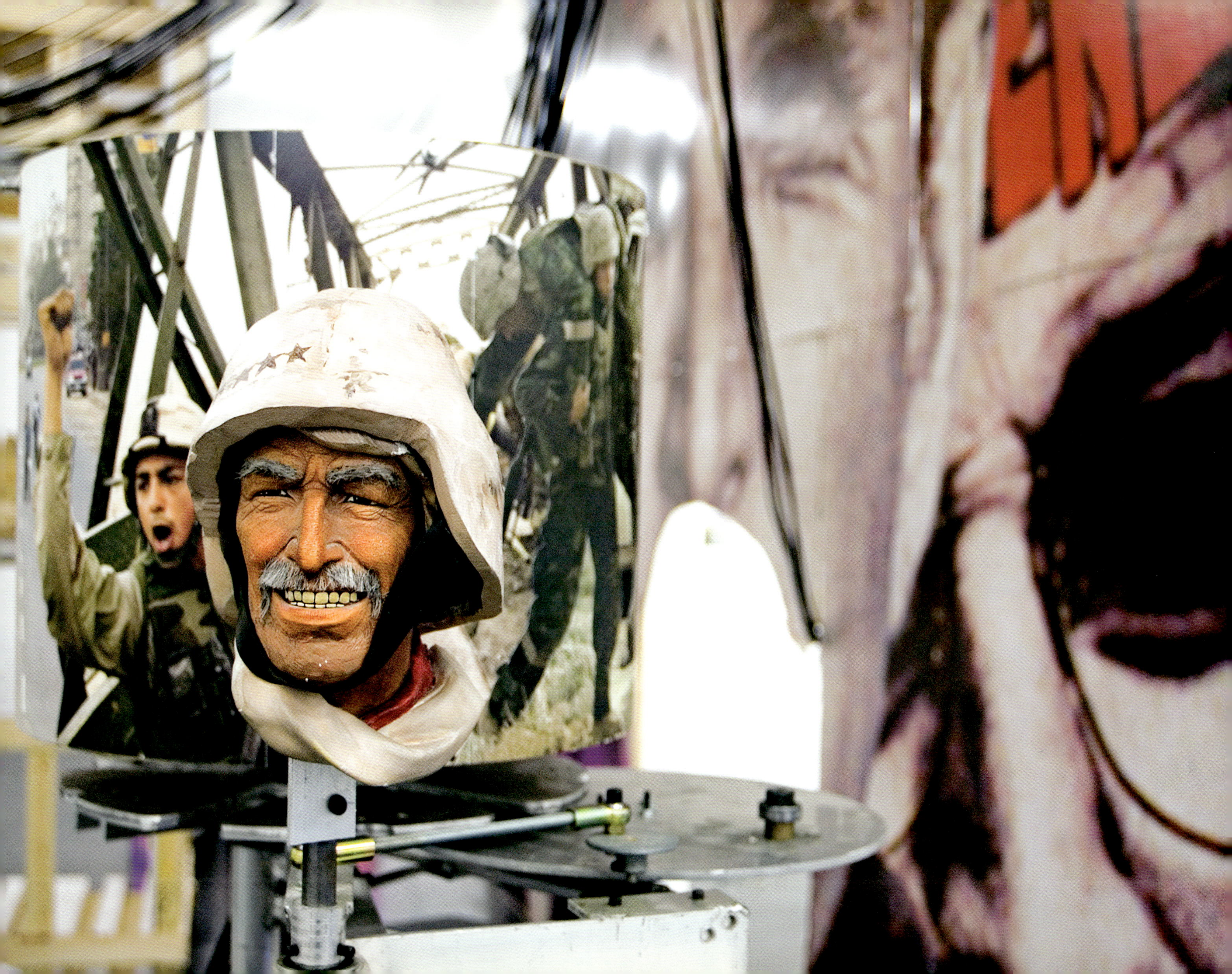

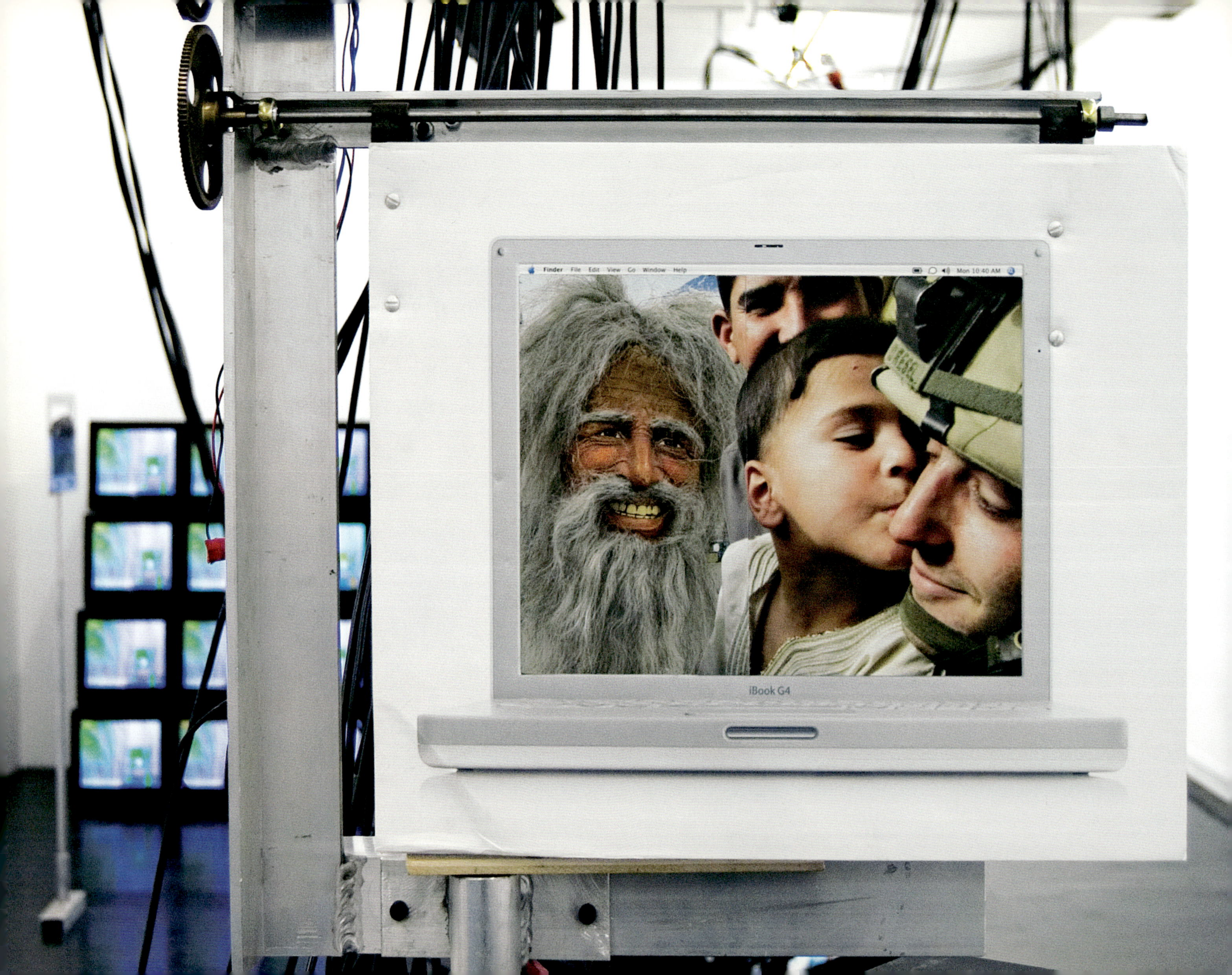
Finder File Edit View Go Window Help
Mon 10:40 AM
iBook G4

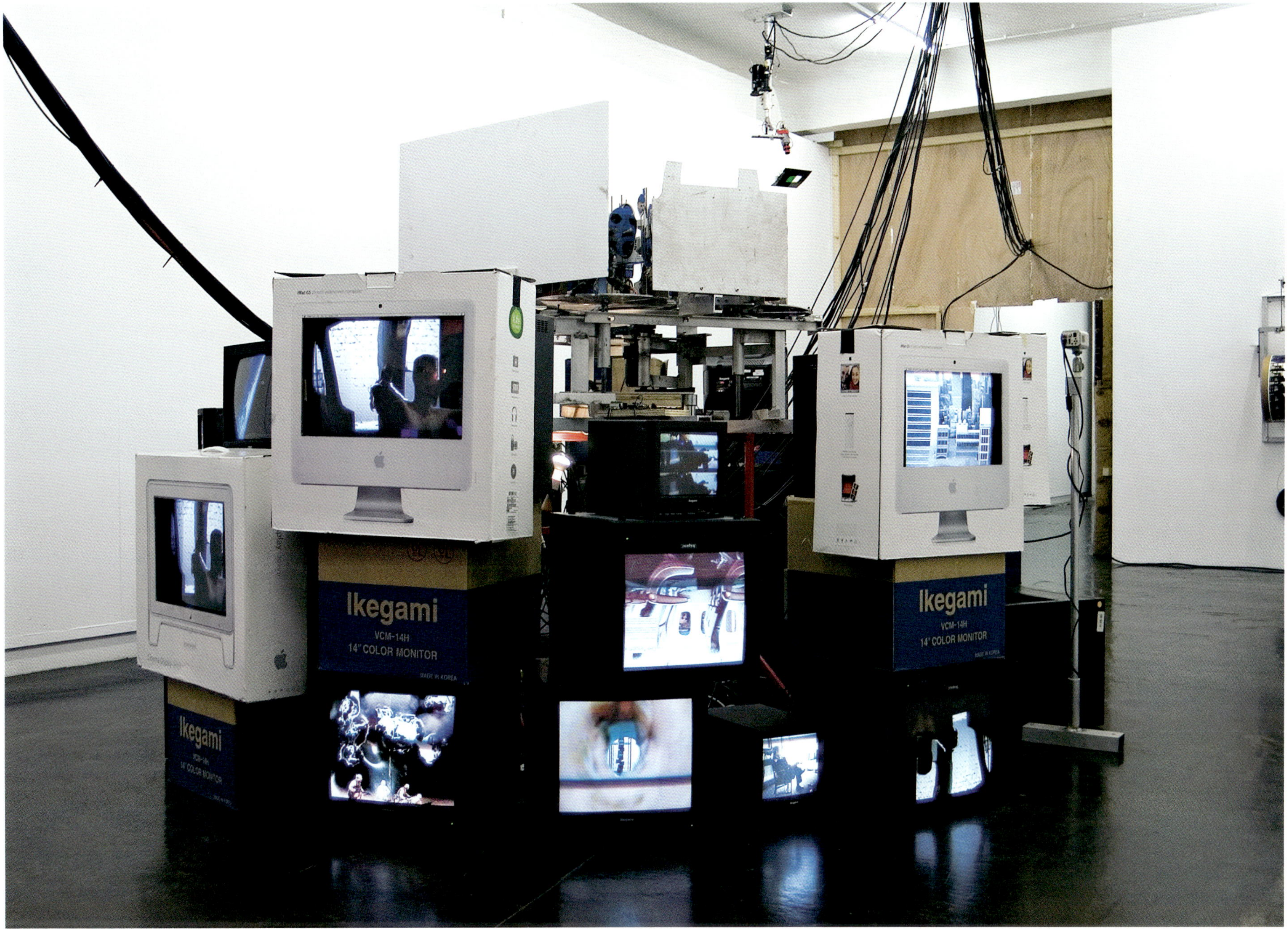
Ikegami
VCM-14H
14" COLOR MONITOR
MADE IN KOREA
Ikegami
VCM-14H
14" COLOR MONITOR
MADE IN KOREA

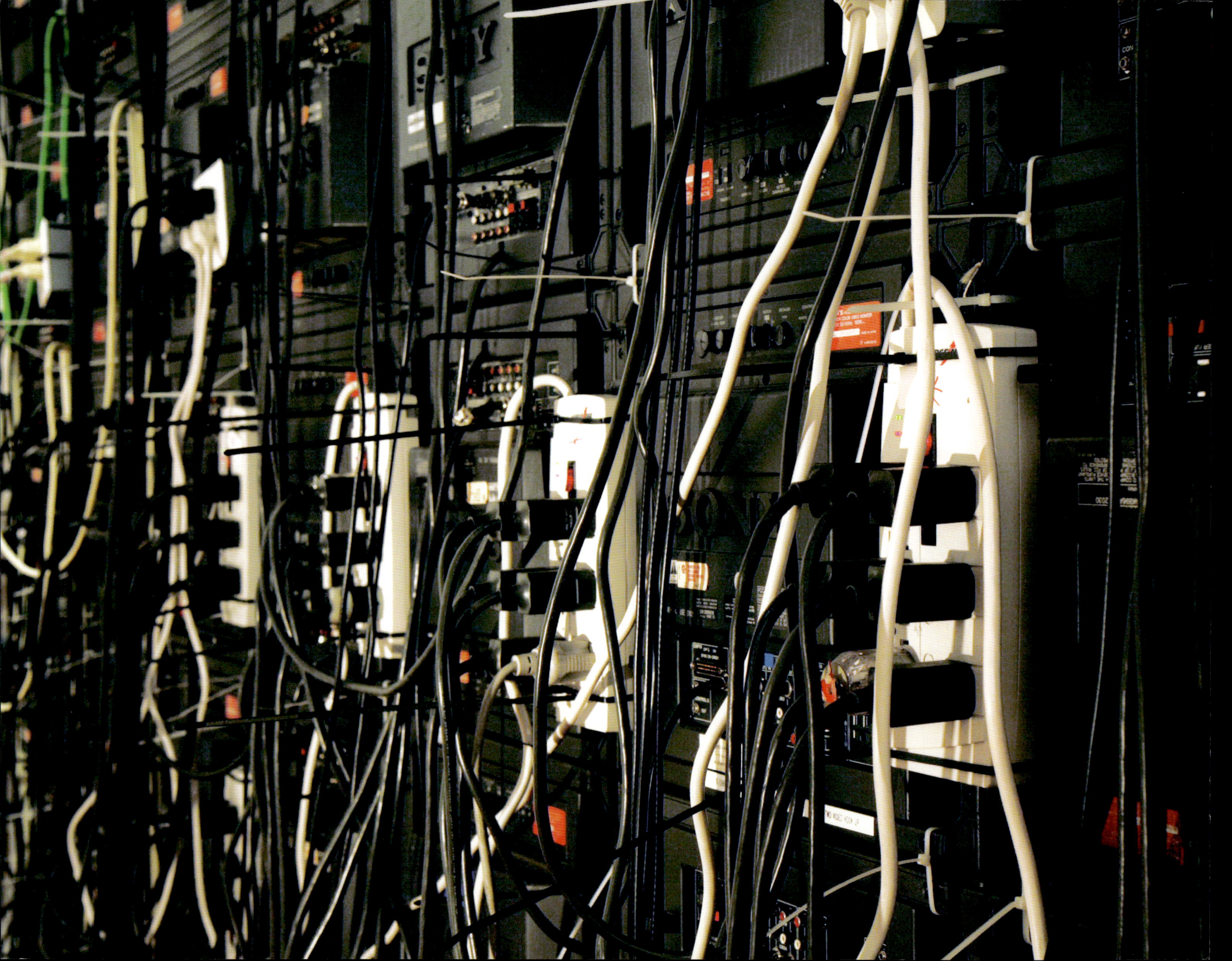

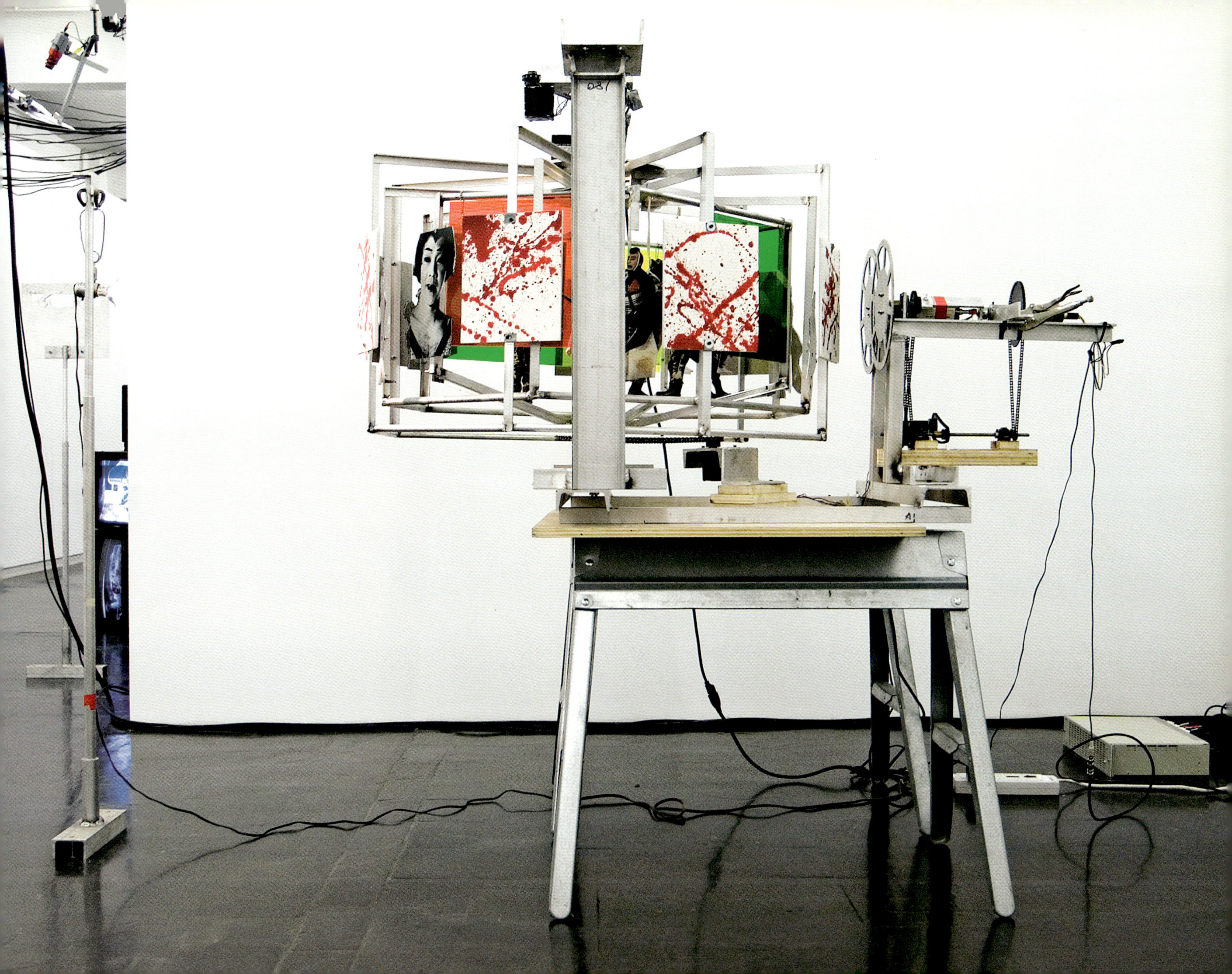

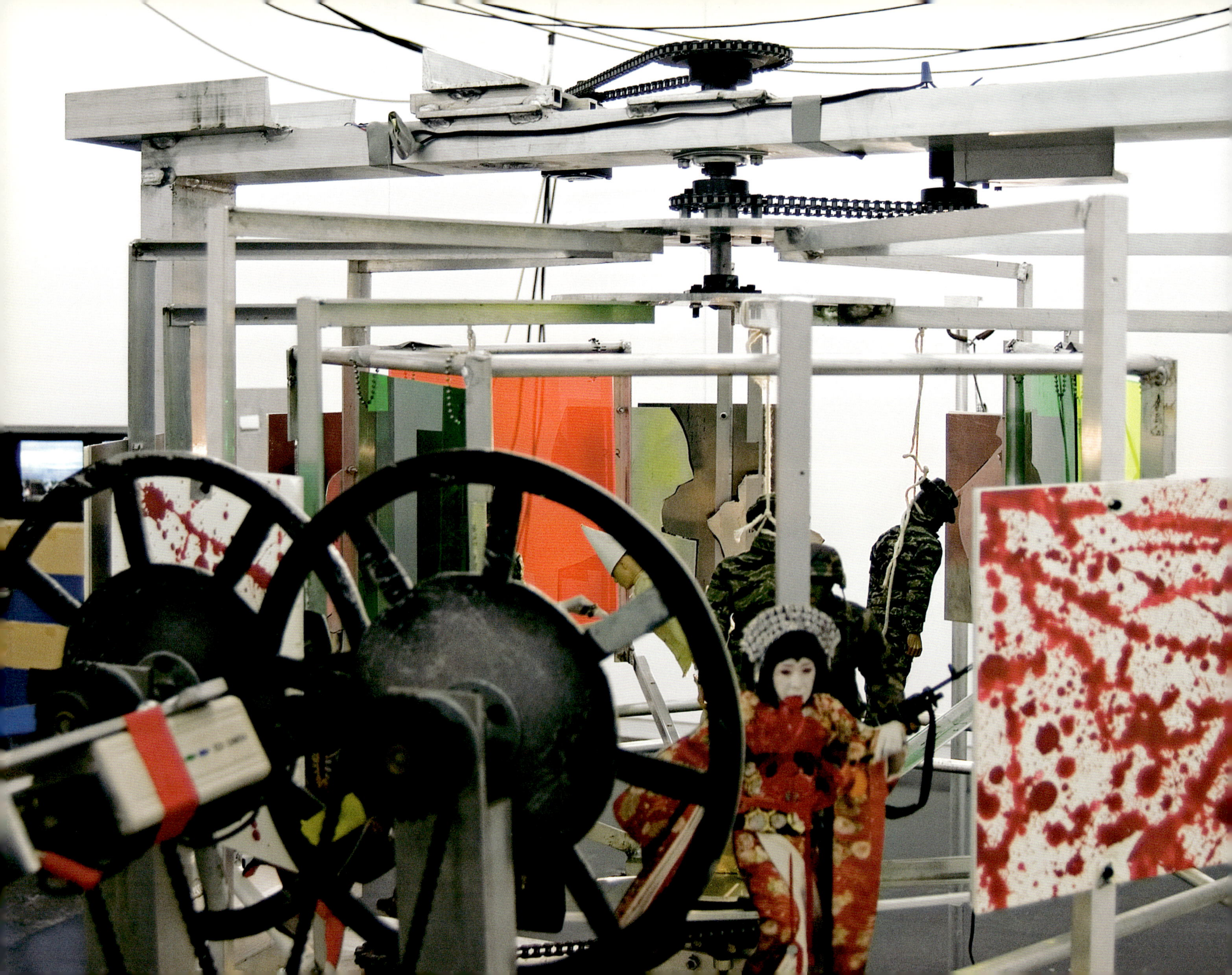

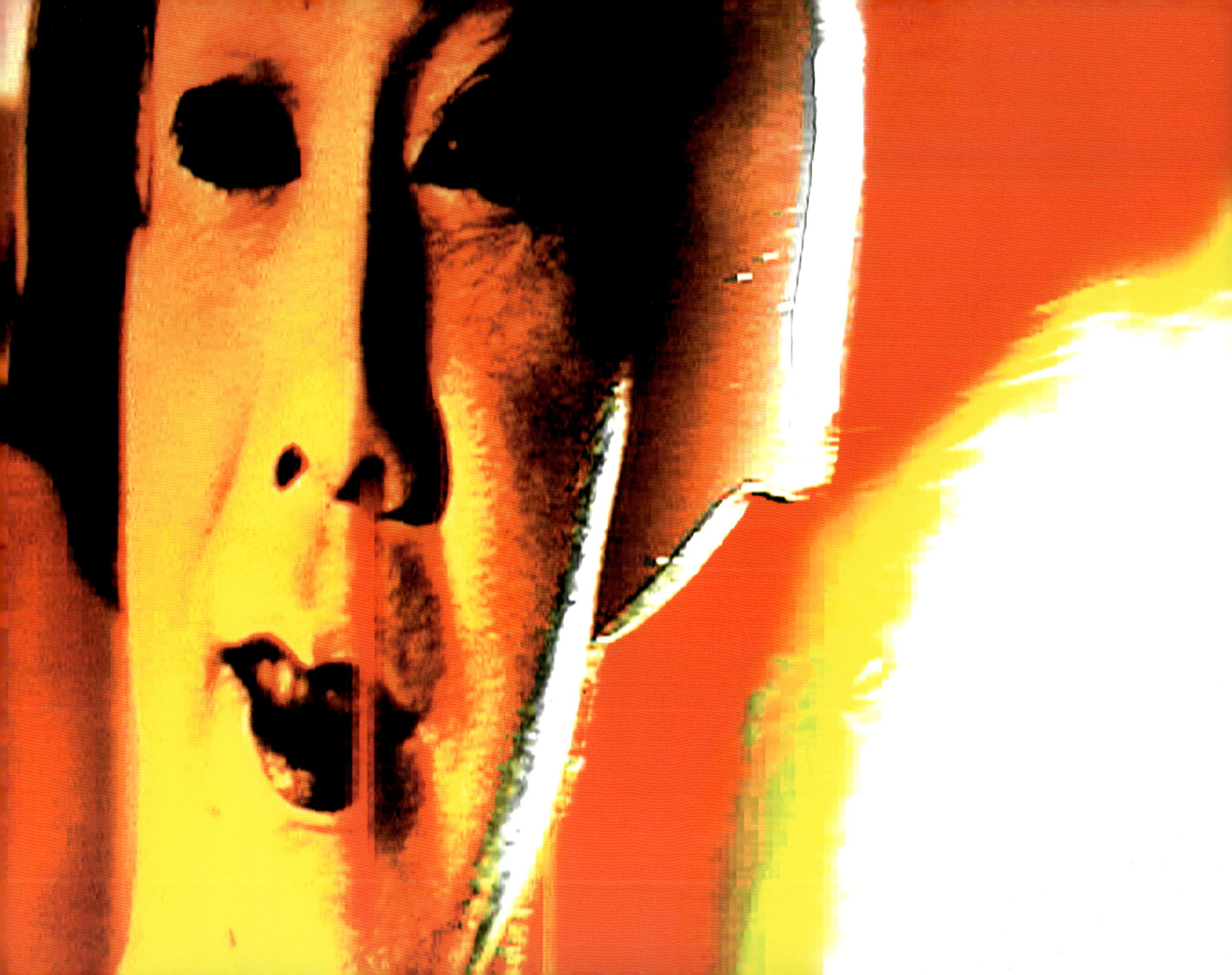

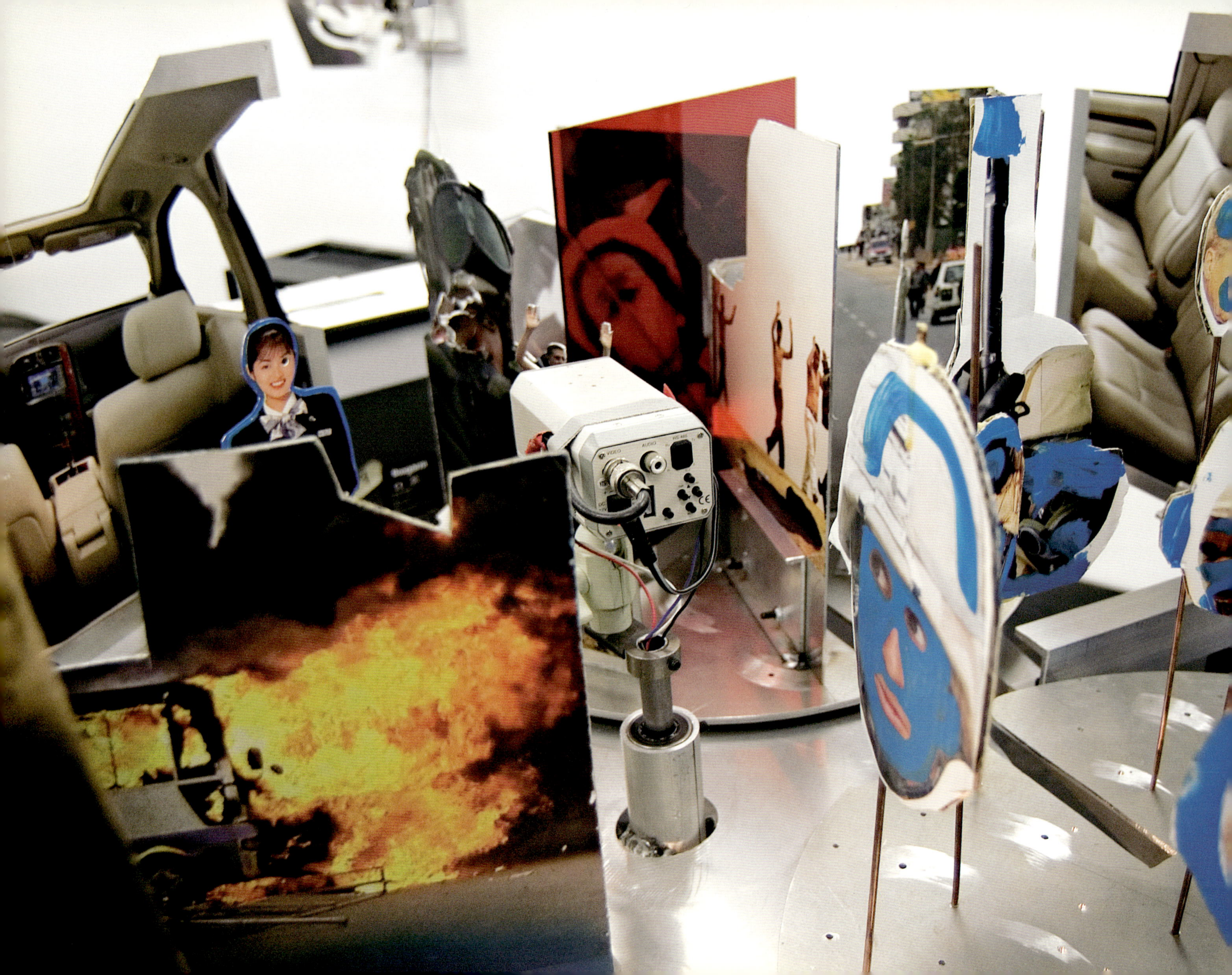

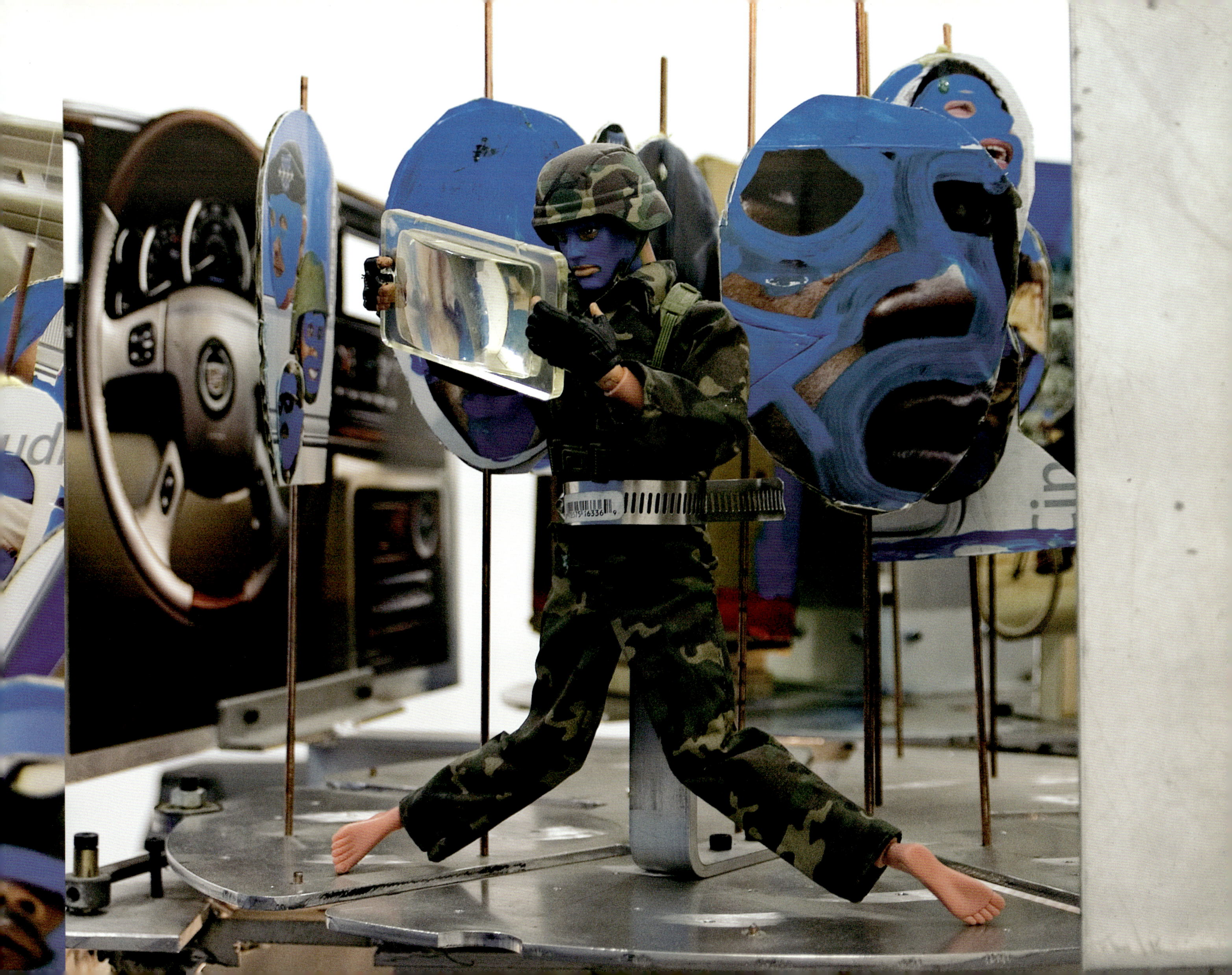

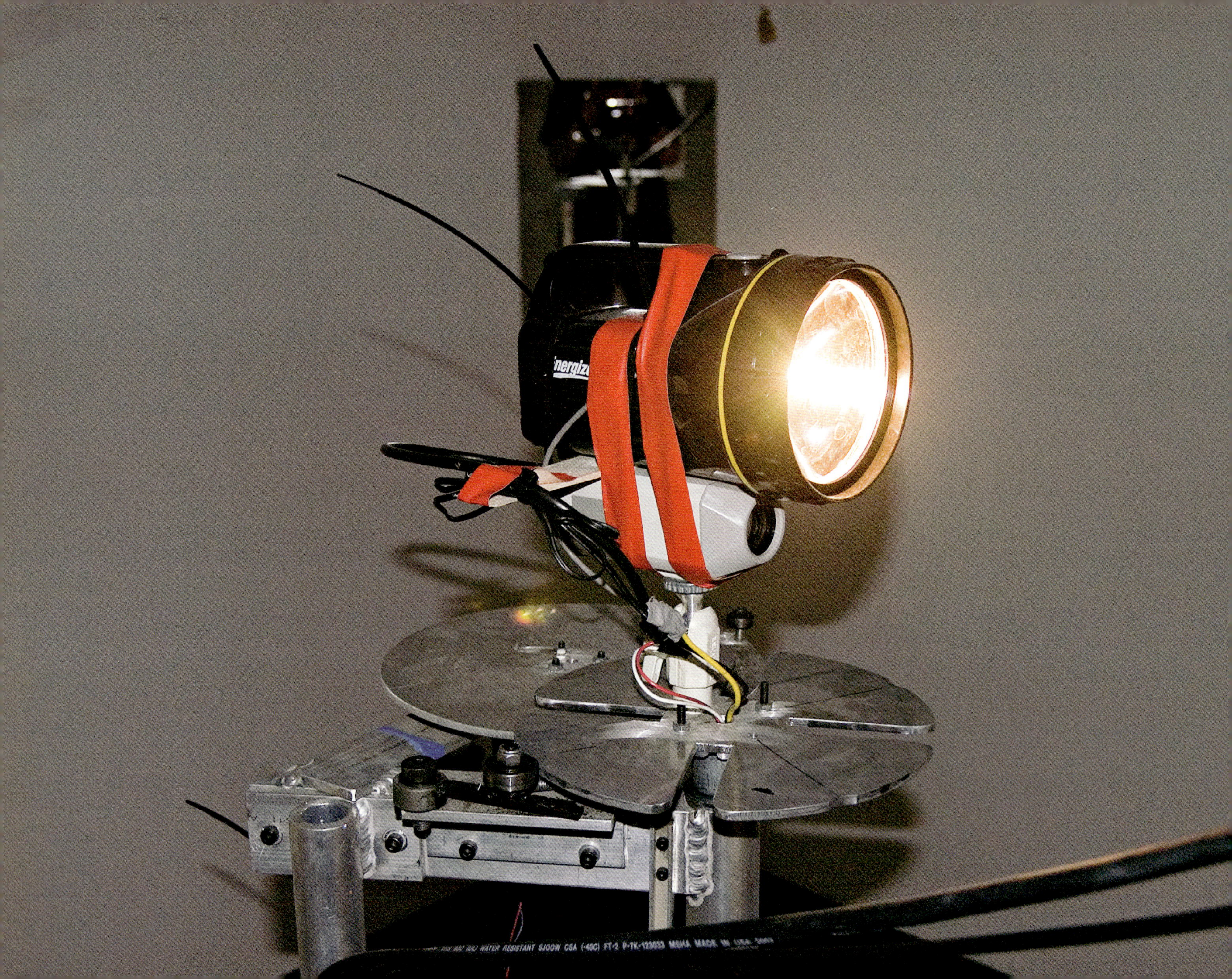

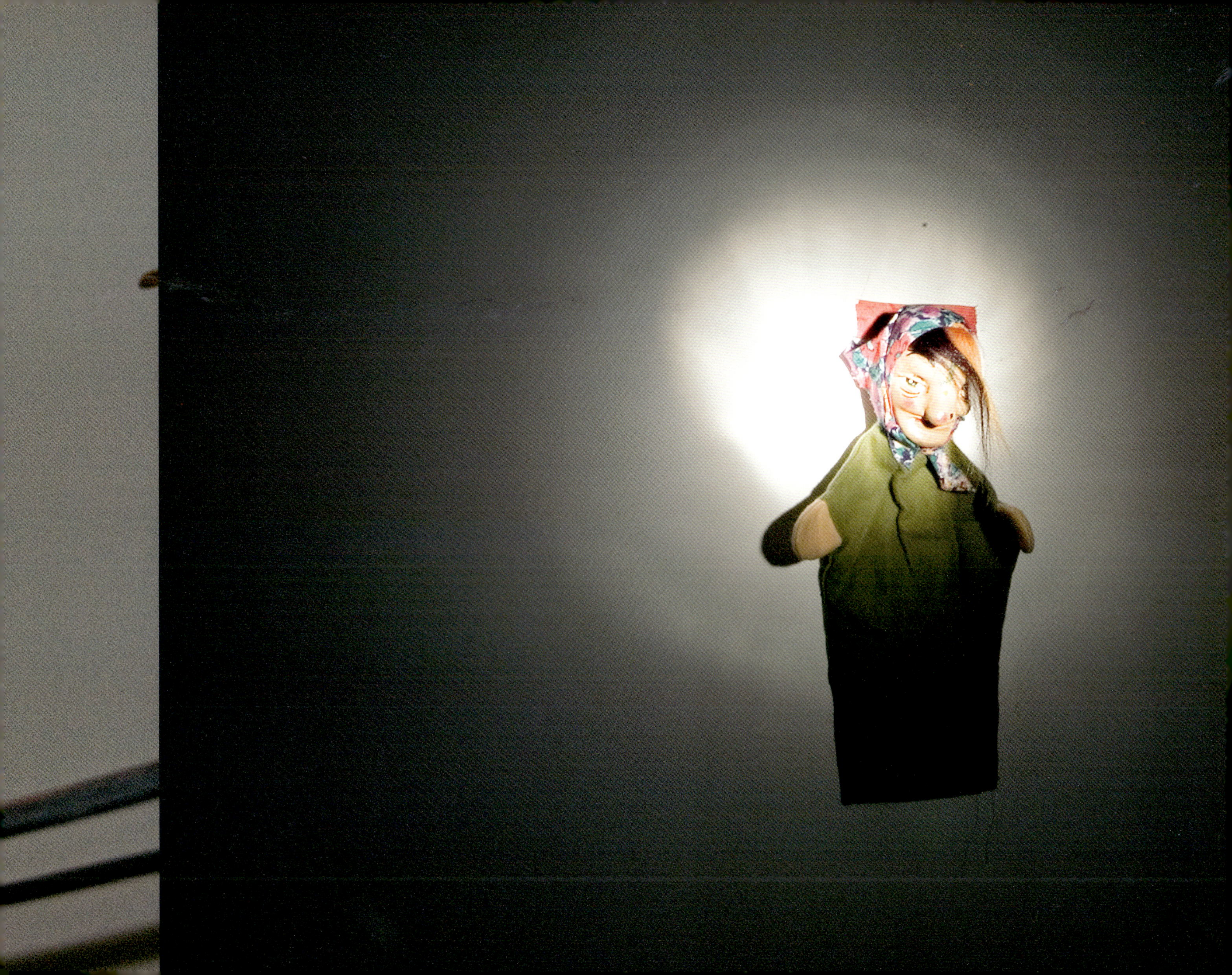

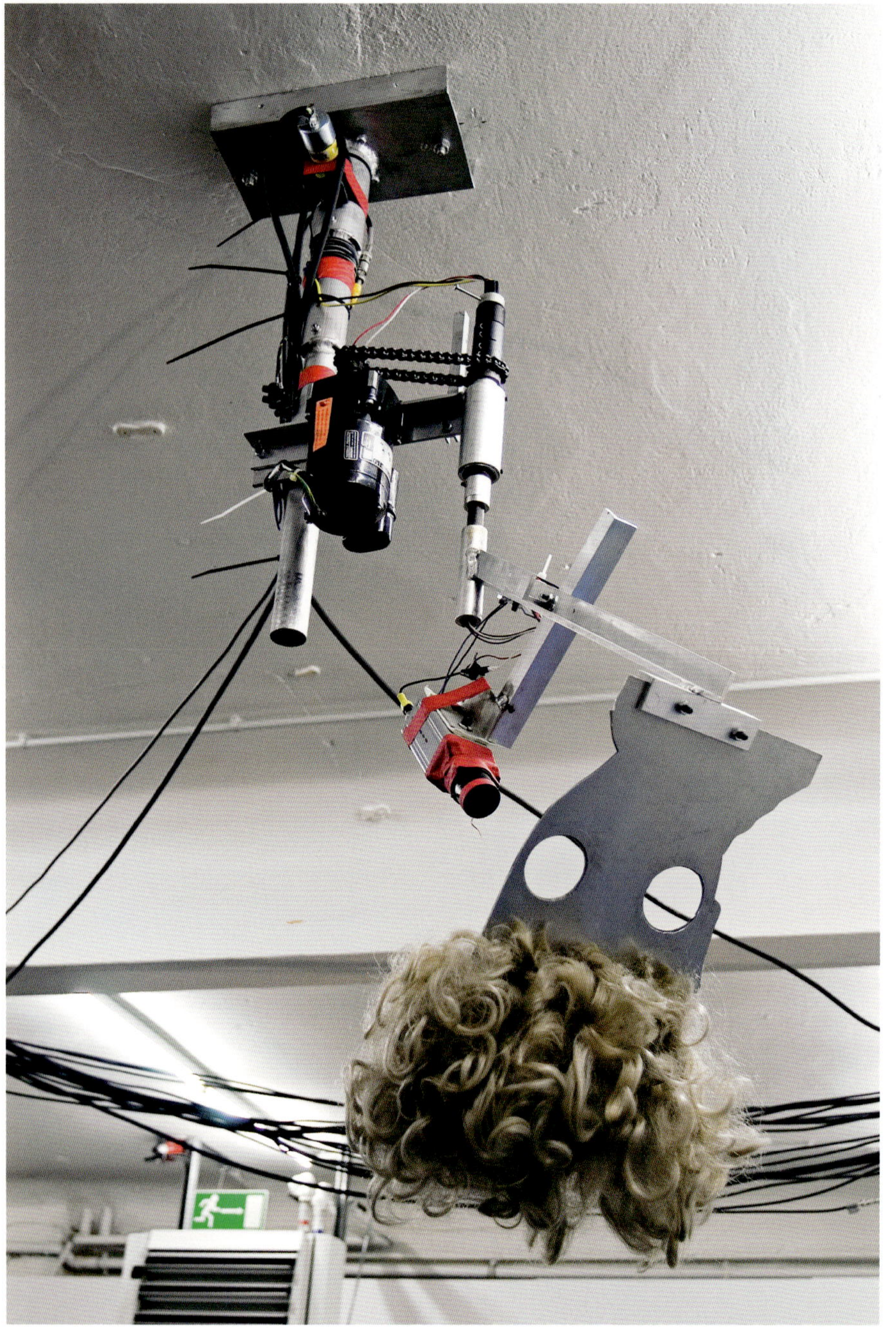

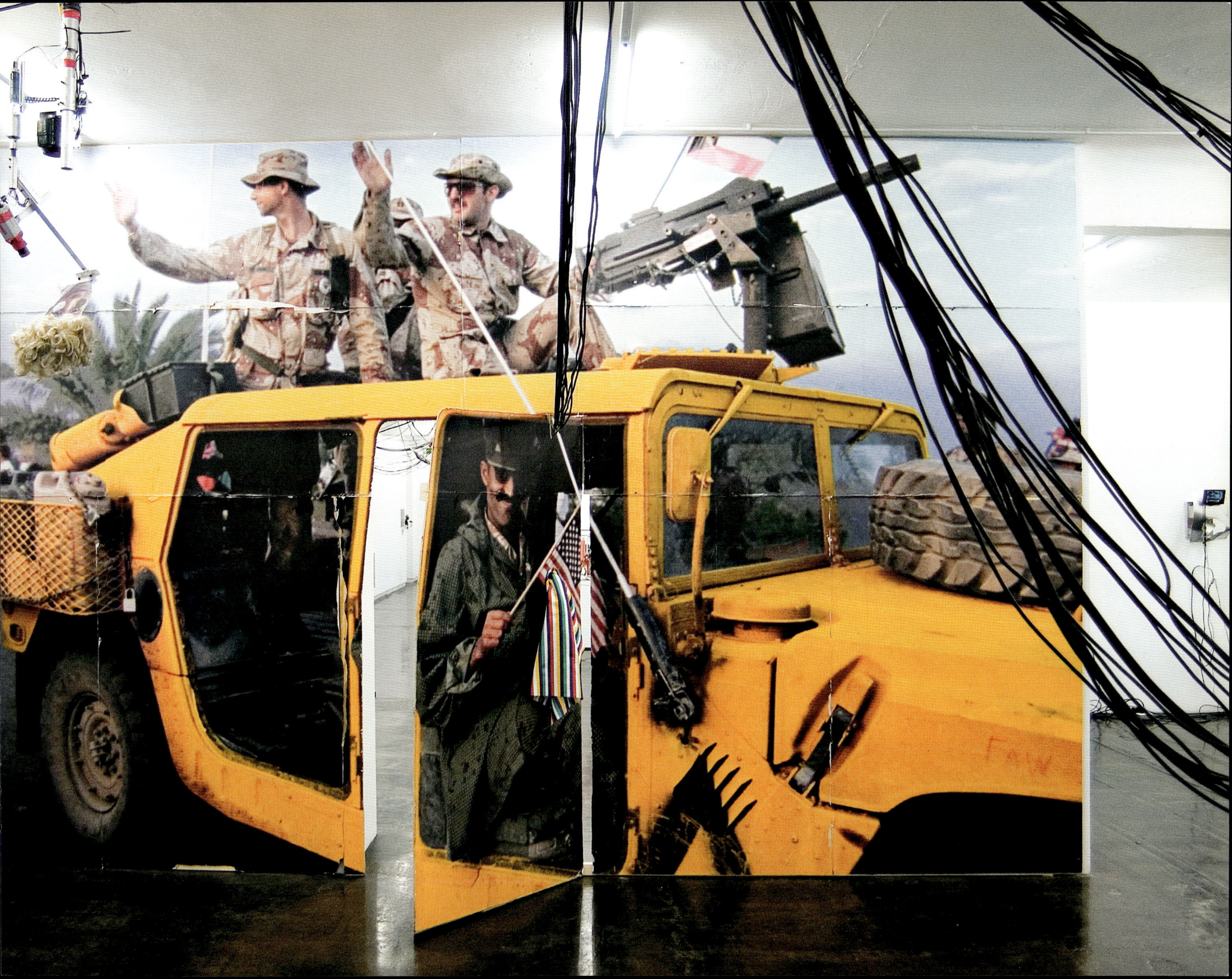

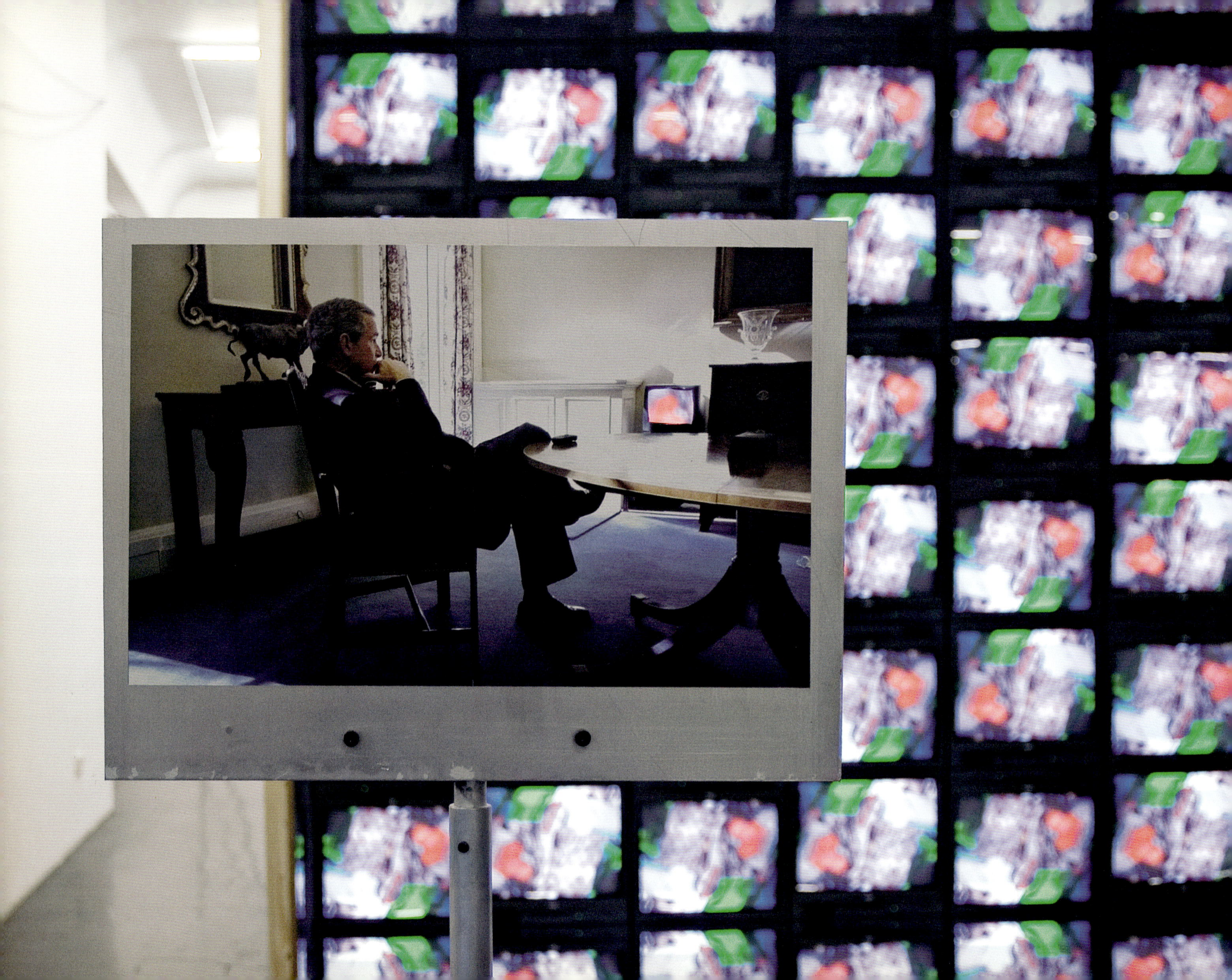

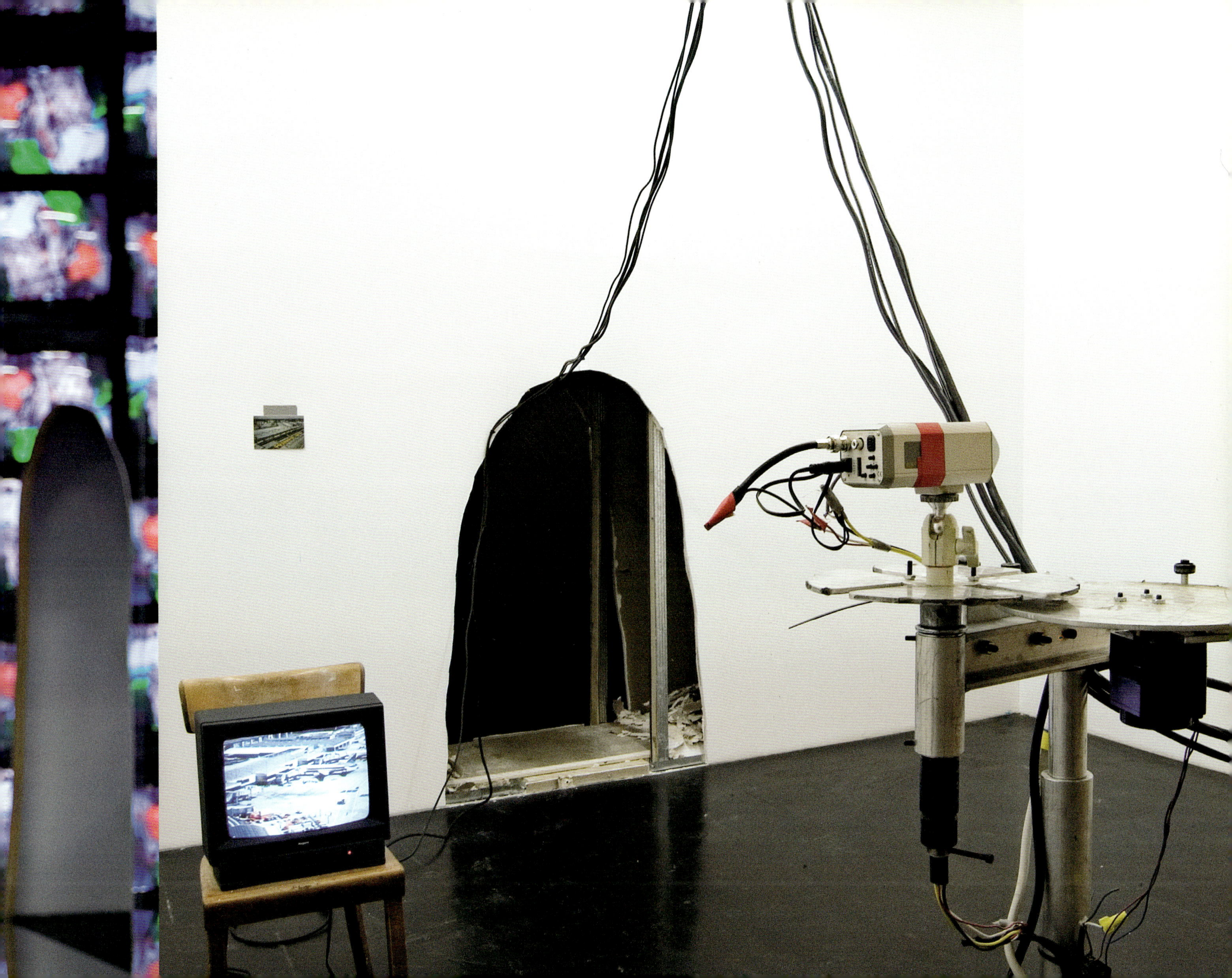

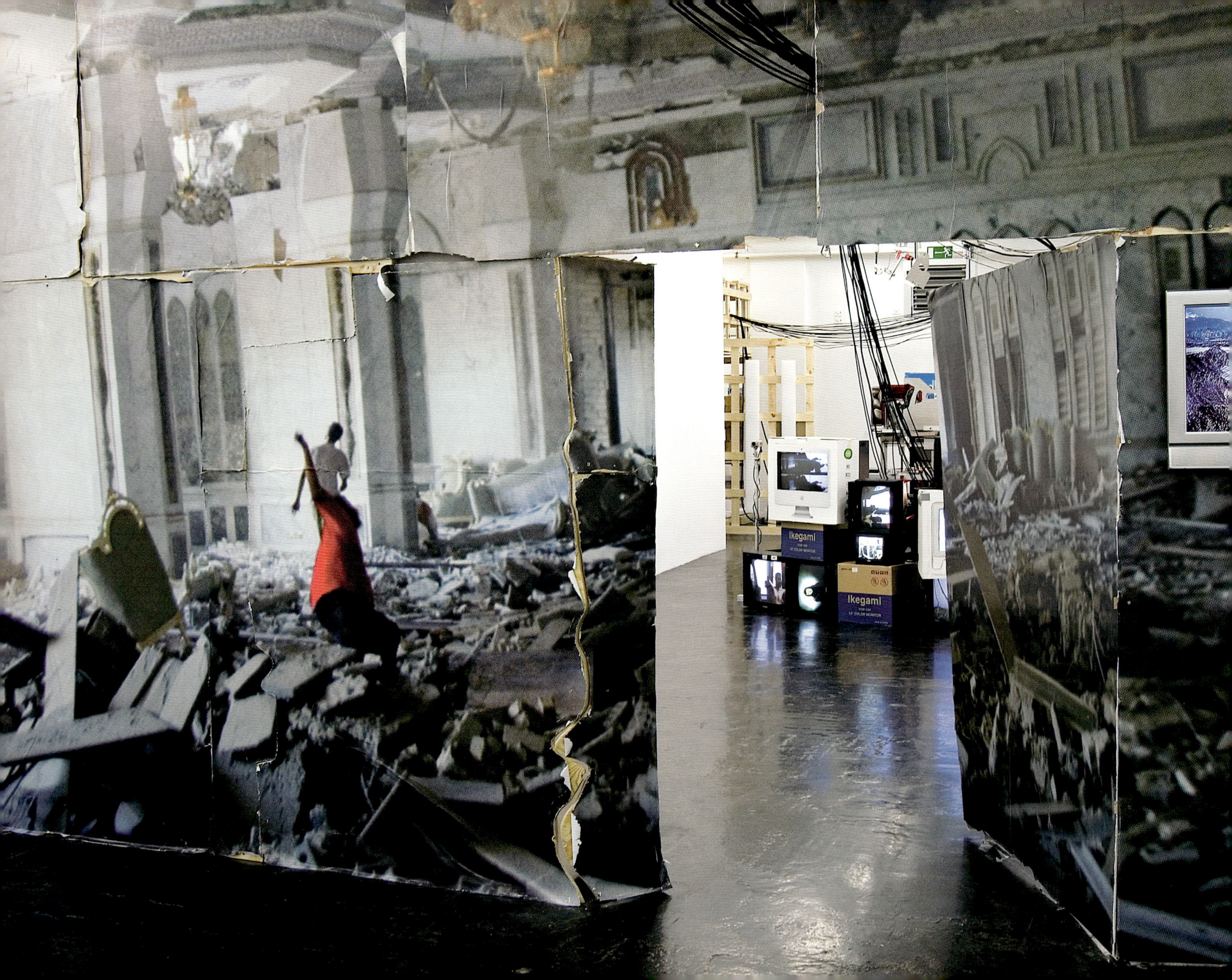
Ikegami
Ikegami

The Palace at 4 A.M.

Alanna Heiss
Director P.S.1 Contemporary Art Center, New York

Jon Kessler, typischer Vertreter der Baby-Boomer-TV-Generation mit ihrer eigenen Bilderwelt, ist dadurch bekannt geworden, dass er Objekte und Abbildungen mit kulturellem Wiedererkennungswert aus ihren kommerziellen Zusammenhängen herauslöst und zu kinetischen Skulpturen verarbeitet. Die Karriere Kesslers begann Mitte der 80er Jahre mit einer Vielzahl von Ausstellungen in den Vereinigten Staaten, Europa und Asien, unter anderem 1984 im „International Survey of Recent Painting and Sculpture" des Museum of Modern Art, 1985 auf der Whitney Biennale und 1986 im „Endgame" des ICA in Boston. Eine der frühen Arbeiten des Künstlers wurde 1984 hier im Project Room des P.S.1 gezeigt.

2004 präsentierte Kessler in New York bei Deitch Projects die Ausstellung „Global Village Idiot". Die mechanisch gesteuerten Skulpturen dieser Ausstellung spiegelten die Stimmung nach 9/11 wider, eine zwiespältige Realität des Lebens zwischen Angst, Fragilität und Terror.

„The Palace at 4 A.M." baut auf dem Projekt „Global Village Idiot" auf, erweitert dieses erheblich und ist die bisher größte und komplexeste Arbeit Kesslers. Der Titel bezieht sich auf eine gleichnamige Arbeit Alberto Giacomettis aus dem Jahre 1932, in der es nach den Worten des Künstlers um die „ungewisse Stimmung zu einer Zeit, in der niemand beobachtet", geht. Die Quellen der Inspiration und der Information – wie Kabuki, Francis Ford Coppolas „Apocalyse Now" und TV-Reality-Shows am Beispiel „The Swan" – variieren erheblich. Letztlich bezieht sich Kesslers Arbeit auf die Geschichte der Repräsentation von Kriegshandlungen und die globale Produktion von Bildern.

„The Palace at 4 A.M." umfasst ein Netzwerk kinetischer Skulpturen mit Überwachungskameras, die im Tandem ihrer Bewegungen Bilder erzeugen, die Videos gleichen. Im Zentrum steht ein Feuerwerk visueller Stimulationen. Auch in den peripheren Bereichen werden Aktionen und Images wiedergegeben, die auf die Abläufe der Monitore und Bildschirme des zentralen „Nervensystems" abgestimmt sind. Die übermittelten Informationen werden fortlaufend aufgenommen und neu zusammengestellt, um so Erscheinungsbilder des Unüblichen, der Desorientierung und nicht zuletzt der Bedrohung zu erzeugen.

Anliegen von Kessler ist die kompromisslose Bloßstellung der Apparate, die er für den Aufbau seiner Bildwelt einsetzt. Die Drähte, Triebwerke, Kameras und Motoren, die für die mechanische Umsetzung sorgen, sind sichtbar und bilden einen Teil des Kunstwerks, indem sie das Innere und Äußere und das Gesehene und Unsichtbare zur Unterscheidung bringen. Aufgrund dieser physikalischen Struktur erscheint „The Palace at 4 A.M." weitgehend demystifiziert, aber doch geben uns die Bilder Rätsel auf.

Kesslers große, vielteilige Skulptur liefert ein Zeitzeugnis aus dem Blickwinkel eines bedeutenden Künstlers. Ich bin stolz darauf, eine Ausstellung mit ebenso schönen wie tragischen Dokumenten unserer Gesellschaft mitverantwortet zu haben.

Pandämonium

Hal Foster

Kein Künstler hat auf die Ereignisse des 11. Septembers und das Debakel des „Kriegs gegen den Terror" pointiertere und kraftvollere Antworten gefunden als Jon Kessler. Mit Bildapparaten, die zugleich komisch, todernst und anarchisch sind, nimmt sich Kessler Stück für Stück das Bush-Regime mit seinen Symbolen einer autoritären Politik, absurden Kriegsführung und beschränkten Kulturvorstellung vor, kaut es durch und spuckt es – als eine Art kalkulierten Zorns gegen das US-Empire – wieder aus.

Kessler ist weithin bekannt für seine Bastelkonstruktionen, in denen er mit Licht- und Bewegungseffekten aller Art Objekte der Massenkultur verarbeitet. In den letzten Jahren hat in seinem Werk eine Verschiebung stattgefunden. Er hat begonnen, mit Videos zu arbeiten, meist in der Low-Tech-Form kleiner Überwachungskameras, die Handlungen von außen und die bizarren Bewegungen kinetischer Objekte auf Monitore übertragen. Er hat das Format seiner mechanischen Tableaus erweitert und die Geräte, Bildschirme, Kabel und Drähte als integrale Bestandteile in seine Arbeit einbezogen. Kessler kommentiert die Bildwelt der Bush-Ära, indem er Nachrichtenschnipsel, Militärberichte, Postkarten, verführerische Kleinanzeigen und Spielzeug zu aberwitzigen Mini-Dramen verarbeitet, die die gegenwärtigen kulturellen und politischen Fixierungen des zeitgenössischen Amerika dekonstruieren. Man stelle sich Moholy-Nagys *Licht-Raum-Modulator* mit Objekten vor, wie man sie auf der *Canal Street* findet, von einem Künstler, der seit Jahren (vergleichbar mit Malcolm McDowell in *A Clockwork Orange*) gezwungen ist, sich die fürchterlichen Ereignisse der Bush-Ära im Fernsehen anzuschauen und zu verarbeiten. Ergebnis dieser Reizüberflutung sind Kesslers Installationen „Global Village Idiot" (2004) und „The Palace at 4 A.M." (2005/06).

Die Zusammenstellung der Bildmaschinen in „Global Village Idiot", die Kessler bei Deitch Projects zeigte, deutet die Zusammenhänge noch diskret an. „The Palace at 4 A.M.", das einige der Stationen von „Global Village Idiot" aufnimmt, ist in seinen Bezügen sehr viel deutlicher und bietet ein weit gefächertes Environment mit vielfachen Brüchen. Zwei der provokativsten Elemente der Installationen sind *One Hour Photo* und *Heaven's Gate*. In *One Hour Photo* rotieren Ramschpostkarten vom World Trade Center auf einem vertikalen Förderband, an dessen unterem Teil eine kleine Videokamera platziert ist. Jedes Mal, wenn eine der Karten sich der Kamera nähert und über sie hinwegkippt, entsteht auf einem Monitor ein verwackelter Zoom in Richtung Twin Towers. Diese Perspektive, die einen dazu zwingt, Teil der Arbeit und damit auch der Attacke auf das World Trade Center zu werden, ist seit dem ersten Golfkrieg in Gestalt kleiner Kameras bekannt, die in die *smart bombs* eingebaut sind. In *One Hour Photo* zeigt Kessler die finalen Bilder der Terroristen und der Opfer an Bord der dem Schicksal geweihten Flugzeuge unmittelbar vor dem Aufprall – Aufnahmen, die in den öffentlich ausgestrahlten Videos ausgeblendet blieben. Kessler, der nur einige Straßenblöcke von Ground Zero entfernt lebt, baut diese traumatischen Bildabläufe in seine Installationen ein, die – wie Freudianer sagen würden – noch nicht verarbeitet sind und deswegen zwanghaft wiederholt werden müssen. Auf diese Weise fängt Kessler einen wesentlichen Aspekt der Katastrophe ein, albtraumhaft real und doch mediatisiert.

Genauso provokativ ist *Heaven's Gate*. Das Video an dieser Station zeigt den Flug durch eine Modellstadt in eine Modellwohnung, in der es auf einen Computerbildschirm zoomt. Wir sehen den Arsch einer Puppe, dringen in ihn ein, nur um auf der anderen Seite in den Galerieräumen wieder aufzutauchen – eine anale Wiedergeburt in die Kunstwelt. Kessler „imaginiert die Repräsentation als einen fleischlichen Akt", wie David Joselit diese Arbeit kommentiert hat. Im Gegensatz zum allgegenwärtigen Mythos virtueller Information werden wir aufgefordert, unseren Blick auf die materiellen und körperlichen Auswirkungen zu richten (Joselit bemerkt dazu, dass „Global Village Idiot" während der Berichte über die Folter in Abu Ghraib auftauchte). Auf einer allgemeineren Ebene werden wir dazu gebracht, uns generell Gedanken über die Obszönität zu machen, die als körperliche Gewalt die Nachrichten in den Medien und unsere Unterhaltungskultur prägt.

Kesslers Arbeit baut auf der Aneignung und Verdrehung dieser Zusammenhänge auf. Sein Titel „Global Village Idiot" ist eine witzige Kombination aus „global village" und „village idiot". Der zweite Begriff bedarf keiner näheren Erläuterung, der erste entstammt dem Werk Marshall McLuhans, der in Texten wie *War and Peace in the Global Village* (1968) den Standpunkt vertritt, dass die Massenkommunikation zum ersten Mal in der Geschichte ein globales Publikum ansprach. Kessler will mit seiner Arbeit sagen, dass man uns im heutigen *global village*, einem von nur wenigen Ländern und Konzernen beherrschten Ort endlosen Infotainments, zu kulturell ignoranten und politisch inkompetenten *village idiots* konditioniert. Die Einsicht, dass technologischer Fortschritt und soziale Regression sich bedingen und nicht Gegensätze sind, hat eine Schlüsselfunktion in seinem Werk. Kessler ist zugleich Idiot und Betrachter. Bush ist es ebenfalls (man wird an den Stoßstangenaufkleber „Somewhere in Texas a Village is Missing its Idiot" erinnert). Und auch Bin Laden ist dabei (in einem Interview stellt Kessler sich die komische Situation vor, dass Osama, der sich irgendwo in einer Höhle verkrochen hat, *Nip/Tuck* anschaut, eine Fernsehserie über kosmetische Chirurgie). So gesehen ist „global village idiocy" ein Zustand allgemeiner Chancengleichheit, gegen den niemand immun ist.

Der Einsatz von Videos, so Kessler, habe ihn befreit. Er könne jetzt Bildabläufe in ihrer Zirkularität und ihren Brüchen darstellen. In „The Palace at 4 A.M." produziert Kessler nicht Bilder, sondern er verschafft den Betrachtern einen Blick hinter die Kulissen auf die Entstehung, die unmittelbar vor dem zerstörerischen Akt angesiedelt ist. Die Mechanismen der Darstellung sind weit entfernt von Perfektion oder Stabilität. Als wären sie Monster eines kleinen Frankenstein, drohen die Bilder, sich gegen ihren Schöpfer, aber auch den Zuschauer zu wenden. Der Rezipient kann sich nicht sicher fühlen. Er sieht nicht nur, sondern er wird auch gesehen, ohne dass jemals zwei Betrachter Zeugen desselben Geschehens sind. Die Maschine und das Bild streben danach, „einander wechselseitig zu vervollständigen, aber das ist am Ende nicht möglich", kommentiert Kessler. Es ginge ihm vielmehr darum, „einen Schnitt" zwischen dem Realen und seiner Repräsentation herzustellen – einen Schnitt, der es uns in die Lage versetzt, die Täuschungsszenarien zu durchschauen. Mit seinen disfunktionalen Installationen will Kessler zeigen, dass sich das große Spektakel amerikanischer Macht in Auflösung befindet, dass die Zauberer und Flüsterer ihr Theater der Illusionen nicht länger aufrecht erhalten können und dass angeblich Wunder verheißende neue Technologien durch immer neue Pannen eingeholt werden. Er weist auf einen zentralen Widerspruch des US-Empire hin: Trotz aller Macht, die weder seine Alliierten noch seine Feinde in Frage stellen, kommt es mit den lokalen Konflikten, vor allem im Mittleren Osten, nicht zurecht. Der Imageverlust der Vereinigten Staaten unter Bush ist gewaltig.

Die von Kessler in „The Palace at 4 A.M." beschriebene Hölle betritt man durch eine Öffnung, die Teil eines gewaltigen weiblichen Geschlechtsorgans ist. („Ich möchte, dass die Zuschauer wiedergeboren werden, wenn sie meine Ausstellung betreten", sagt Kessler, „aber auf eine unreine Weise.") Der erste Blick wandert auf eine Fotowand mit einer zertrümmerten Residenz, ehemals Palast von Saddam Hussein. Zugleich sieht man eine großformatige Fotografie von Bush, auf die in Blutrot das Wort „War" gekrakelt ist. Der Palast könnte ebensogut das *Weiße Haus* sein oder ein Psychobunker, den sich Bush und Hussein miteinander teilen und in den via Medien aber auch wir eingeladen sind. Auf den ersten Blick hat dieses *Gesamtkunstwerk* den karnevalesken Charakter einer verkehrten Welt, in der Oben und Unten, Innen und Außen vertauscht sind. Es geht dabei auch um uns selbst und unsere Albträume tief in der Nacht, die sich mit dem „Krieg gegen den Terror" befassen und von Gewalt geprägt sind.

Mit seiner Kunst bezieht sich Kessler auf zahlreiche Vorgänger: auf Robert Rauschenberg und seine wilden Kombinationen von Medienzitaten, auf Claes Oldenburg und sein regressives Theater von Bastelobjekten, auf Jean Tinguely und seine autodestruktiven Apparate und viele mehr. Gegenwartsnah ließe sich an Mike Kelley denken und seine skurrilen Inszenierungen. Oder an Medientheoretiker wie Paul Virilio, an Filmemacher wie David Cronenberg („The Palace at 4 A.M." könnte auch den Titel „Videodrome" tragen) und an Schriftsteller wie Thomas Pynchon oder Philip K. Dick. In einem spezifischen Zusammenhang steht „The Palace at 4 A.M." mit Alberto Giacometti und dessen surrealistischen Spielbrettern und Käfigen (Kesslers Titel ist einer Arbeit Giacomettis aus dieser Werkphase entliehen). Wie die Surrealisten verfolgt Kessler die Strategie, Bilder und Objekte der Alltagskultur zu einer höheren Schönheit zusammenzuführen, in der Begehren und Tod miteinander verknüpft sind. Und es gibt einen weiteren Bezug. Mit seiner Arbeit *Le Palais à 4 heures du matin* liefert Giacometti einen Entwurf zur Frage ödipaler Subjektwerdung. Dies ist auch ein Anliegen Kesslers, der mit der Generation der Väter abrechnet und die Frage stellt, wie Menschen unter dem Regime globaler Unterhaltung und imperialer Politik zurechtkommen.

Insofern zieht sich wie im Surrealismus und in den Werken von McLuhan, Virilio, Cronenberg, Pynchon und Dick ein Element der Paranoia – kritischer Paranoia – durch die Arbeit Kesslers. Dick hat den Paranoiker als eine Person gesehen, die im Besitz aller Fakten ist, wohingegen Freud den Paranoiker als ein Subjekt verstand, das sich verzweifelt bemüht, nicht kohärente Fakten notfalls über Verschwörungstheorien miteinander zu verknüpfen. Freuds Kollege Victor Tausk konzentriert sich in einer Studie auf Paranoiker, die sich von „Beeinflussungsapparaten" leiten lassen. Auch Kessler, der in seiner Arbeit zwischen Verbindendem und Trennendem laviert, nimmt immer wieder Bezug auf die Interessengruppen in unserer Gesellschaft mit ihren „Beeinflussungsapparaten" (vielleicht auch das ein treffender Titel seiner Arbeit). Und doch weigert sich Kessler, sich den Apparaten zu unterwerfen. Seine Installationen sind so konstruiert, dass sie den Bilderfluss der Machtmaschinerie jederzeit, wenn auch nur kurzfristig, unterbrechen können.

Ich möchte mit einem weiteren Bezug schließen, der meiner Meinung nach weiter reicht als alle anderen. Vor beinahe hundert Jahren, inmitten des Ersten Weltkriegs, entwickelten die Züricher Dadaisten eine Strategie der mimetischen Provokation. In ihrem Cabaret Voltaire nahmen sie die Schlagworte und Bilder aus der Propaganda der sich bekriegenden Weltmächte auf und hinterfragten sie mit bissigem Humor. „Was wir Dada nennen", schrieb Hugo Ball im Manifest des Dada von 1916, „ist ein Narrenspiel aus dem Nichts, in das alle höheren Fragen verwickelt sind; eine Gladiatorengeste; ein Spiel mit den schäbigen Überbleibseln; eine Hinrichtung der posierten Moralität und Fülle." Kessler stellt ein Cabaret Voltaire der Gegenwart vor. Nahm Ball seinerzeit die Rolle des „magischen Bischofs" ein, so setzt sich Kessler heute die Tarnkappe des „globale village idiot" auf – eines Schamanen, der versucht, die bösen Geister aus dem Pandämonium der Teufel, der Gewalt, des Aufruhrs und der Verwirrung zu vertreiben.

Ludwig Seyfarth

Überwachter Spielplatz.

Installationskunst in der Kontrollgesellschaft am Beispiel von Jon Kesslers „The Palace at 4 A.M."

„Wir leben in einem Zeitalter, in dem zum ersten Mal Tausende höchstqualifizierter Individuen einen Beruf daraus gemacht haben, sich in das kollektive öffentliche Denken einzuschalten, um es zu manipulieren, auszubeuten und zu kontrollieren."[1]

Mit diesen Worten leitete Marshall McLuhan 1951 sein erstes medientheoretisches Hauptwerk „The Mechanical Bride" ein, in dessen collageartig aneinander gereihten Kapiteln er vor allem die manipulative Wirkung der Werbung analysierte.

Was McLuhan vor über fünfzig Jahren für die Konsumkultur konstatierte, scheint inzwischen das politische und gesamte gesellschaftliche Leben Amerikas und vieler anderer Nationen erfasst zu haben. Spätestens in der Folge der Anschläge vom 11. September verwandeln sich vermeintlich demokratische Gesellschaften zunehmend in paranoide Überwachungsstaaten, in denen ständig und so viel wie möglich kontrolliert wird. Schon eine leicht auffällige Kopfform gerät durch die Erfassung biometrischer Daten zum verdächtigen Merkmal. Am Mainzer Hauptbahnhof wurde mit Hilfe von 200 freiwilligen Testpersonen von Oktober 2006 bis Januar 2007 ein Video-Überwachungssystem getestet, das Einzelpersonen in großen Menschenmengen auffindbar machen soll. Datenschützer kritisierten unter anderem, dass das System automatisch auch Daten vieler anderer Passanten speicherte.

Schließlich kommt es vor, dass man sich in ungewohnten Situationen selbst nicht mehr erkennt. Man steht vor einem Monitor, ohne zu wissen, dass das Bild von einer Überwachungskamera kommt, und realisiert erst nach einiger Zeit, dass man es selbst ist, der da etwas unmotiviert herumsteht und dabei keine besonders gute Figur abgibt.

Vor einigen Jahren fragte eine Reklametafel in Manhattan: „An jedem durchschnittlichen Tag werden Sie mindestens ein Dutzend Mal von Überwachungskameras gefilmt – sind Sie auch schick genug angezogen?"[2]
Die Werbung fordert uns auf, keine Angst vor der permanenten Kontrolle zu haben, sondern sie ebenso narzisstisch zu goutieren wie Jennifer Ringley, eine junge Amerikanerin, die schon vor zehn Jahren ihr Privatleben über eine Webcam im Internet veröffentlichte und damit vor allem die Neugier auf etwaige Sexszenen weckte. Unzählige Porno-Websites wurden inzwischen nach der „JenniCam" benannt.

„Das Lustprinzip des Voyeurs, alles zu sehen, und das Lustprinzip des Exhibitionisten, alles zu zeigen, werden von privaten Triebschicksalen zu sozialen Normen",[3] schreibt Peter Weibel. Auch der Narzissmus werde von einer individualpsychologischen zu einer sozialen Kategorie, was Marshall McLuhan schon 1964 voraussah, als er ein Kapitel von „Understanding Media" dem Narzissmus als „Narkose" widmete. Er widerspricht der traditionellen Deutung der Selbstverliebtheit des Narziss und stellt in Frage, ob Narziss im Spiegelbild überhaupt sich selbst erkannt habe. McLuhan interpretiert den Narzissmus medientheoretisch als Verliebtheit des Menschen in die von ihm geschaffenen Apparate, als Narkotisierung durch die technischen Ausweitungen des Körpers.[4]

Verliebtheit ins eigene Spiegelbild heißt demnach, sich so mit dem medial übertragenen Bild seiner selbst zu identifizieren, dass man sich nicht mehr von ihm zu unterscheiden vermag. In diesem Zustand ist ebenso wenig ein klarer Blick auf die Verhältnisse möglich wie bei der angstbesetzten Erregung im Zustand moralischer Empörung. Anstatt vor lauter Abscheu gar nicht hinzuschauen, sollte man versuchen, wachen Auges die Mechanismen zu durchschauen, die uns in unterschwelliger Abhängigkeit halten.

Schon in „The Mechanical Bride" hielt McLuhan die moralische Entrüstung für einen schlechten Ratgeber, wenn man im Strudel der Medienbilder nicht hilflos ertrinken will. Aber wie sollte man sich verhalten? McLuhan bedient sich einer literarischen Metapher und verweist auf Edgar Allan Poes Erzählung „Der Sturz in den Malstrom", in der es einem Schiffbrüchigen gelingt, sich in aussichtslos erscheinender Lage vor dem Ertrinken zu retten. Inmitten eines gewaltigen Wasserwirbels, der ihn mit Gewalt herabzuziehen droht, kommt er auf die ihm selbst irrsinnig erscheinende Idee, dass er „eine Art *Vergnügen* darin suchte, Spekulationen über die verschiedenen Fallgeschwindigkeiten anzustellen, mit denen sie dem Gischtpfuhl tief dort unten zustrudelten". Indem der Seemann sich so verhält wie die nach oben treibenden Objekte, bleibt er am Leben. Analog dazu versucht McLuhan in „The Mechanical Bride", „den Leser in den Mittelpunkt eines durch die Kräfte in Rotation versetzten Bildes zu stellen, von wo aus er die Vorgänge beobachten kann, die gerade ablaufen und in die jeder verwickelt ist".[5]

McLuhans Lesart der Poe'schen Erzählung deutet künstlerische Strategien an, die sich erst in den folgenden Jahrzehnten entwickeln sollten und die Rezipienten einem räumlichen Szenario aussetzen, das zunächst als chaotische Überfülle erscheint und erst bei genauerer Beobachtung von Einzelheiten Kontur gewinnt. Es handelt sich um unterschiedlichste Werke, die im weitesten Sinne unter die Kategorie „Installation" fallen. Dass es hier noch schwerer ist, sinnvolle Kategorisierungen, Typisierungen und Gattungsdefinitionen vorzunehmen als in anderen Bereichen der heutigen Kunst, legt es nahe, darauf programmatisch zu verzichten, wie es Juliane Rebentisch in ihrer Abhandlung tut, die das Thema Installation aus Sicht der philosophischen Ästhetik einkreist.[6] Gleichwohl möchte ich grobe Tendenzen unterscheiden, um so auch die Spezifika der Installationskunst Jon Kesslers näher zu umreißen.

Die Tendenz zur chaotisch-übervollen Installation, zu einer schier unersättlichen Anhäufung von Materialien, mag manchmal hypertroph und größenwahnsinnig erscheinen. Ich denke an Künstlerinnen und Künstler, die ich einigermaßen chronologisch von den 1960er Jahren bis heute aufzähle, wie Edward Kienholz, Paul Thek, Anna Oppermann, Paul McCarthy, Mike Kelley, Jason Rhoades, John Bock oder Jonathan Meese.

Wenn ich Jon Kessler mit seinem „Palace at 4 A.M."[7] nicht in diese „Kategorie" subsumiere, dann deshalb, weil hier, trotz der Größe und Fülle, eine andere Tendenz oder Tradition hineinspielt, in der sparsamere und überschaubarere räumliche Szenarien dominieren. Dies sind Installationen, die von einem oft irritierenden Einsatz von Überwachungskameras geprägt sind, den wir auch bei Jon Kessler finden – worauf ich später zurückkomme.

Installationen sind raumgreifend und haben dadurch etwas Theater- und Bühnenhaftes, auch wenn keine theatralische Handlung in Form einer Performance oder eines Happenings stattfindet. Insofern ist eine Installation auch eine Art verräumlichtes Bild, das erst durch die Aktionen der Betrachter in Bewegung gerät. Dies gilt für „The Palace at 4 A.M." in besonders raffinierter Weise, denn hier ist auf einer Art zweiten Ebene auch stillgestellt, was sich offenkundig zu bewegen scheint, nämlich das, was wir auf den unzähligen Monitoren sehen. Der Bilderzufluss ist auf doppelte Weise streng limitiert: erstens werden keine aufgezeichneten Bilder gezeigt (mit der Ausnahme einer computersimulierten Satellitenaufnahme und Landkarte aus dem Irak sowie einer Sequenz, die in einem Satellitenbild von Hamburg-Harburg mündet und den Blick eines Bomberpiloten auf den Ausstellungsort suggeriert. Ein entsprechendes Video wurde auch in der ersten Präsentation der Installation im P.S.1 1 in New York gezeigt); zweitens wird nur auf die Monitore übertragen, was sich innerhalb der Installation befindet (zu deren Gesamtheit auch einzelne Ausblicke aus den Fenstern gehören). Trotz des hohen technischen Aufwandes handelt es sich um eine Apparatur, die nicht wirklich „funktioniert".

Hier besteht eine Vergleichbarkeit zu den Installationen des Belgiers Guillaume Bijl, der 1979 damit begann, Orte des Konsums und andere Räume des sozialen Lebens – allerdings nur als Ausstellungsstück – nachzubauen. In Bijls Fahrschule kann man keinen Führerschein machen, im Supermarkt nicht einkaufen, im Auktionshaus nichts ersteigern und im Fernsehstudio findet keine Quiz-Show statt.[8]

So „realistisch" alles erscheint, handelt es sich bei dem „TV Quiz Decor" von 1993 um eine Attrappe, um eine bloße Fassade. Wer hinter die Kulissen des bunt glitzernden Szenarios schaut, wird kein Aufnahmestudio finden, das irgendeine Fernsehübertragung gewährleisten könnte. Bei Kessler funktioniert zwar die technische Installation, aber es wird nur aufgenommen, was sich innerhalb der Installation befindet. Während es bei Bijl kein Hinter-den-Kulissen gibt, realisiert Kessler gleichsam nur das „Backstage", aber es gibt kein Davor, das hier sinnvoll verarbeitet werden könnte.

Stellt Bijl die sozialen Orte der Konsumgesellschaft als funktionslose Stilleben vor, tut Kessler etwas Analoges mit der Berichterstattung über Krieg und Terrorismus in den Medien. Er führt uns in eine selbstgebastelte gigantische Kriegsspielwiese mit Soldatenfiguren, gestanzten und gebogenen Blechen, auf Sperrholzplatten geklebten Großfotos, paternosterartig rotierenden Ansichtskarten des World Trade Centers, diversen kreisende, rotierende oder abgehackt zitternde Bewegungen vollführende Apparaturen, an denen Bilder, diverse Gegenstände oder Kameras befestigt sind, endlosen oft mitten durch den Raum hängenden Kabelverbindungen und unzähligen aufgehängten, auf Podesten stehenden oder zu Wänden zusammengestellten Monitoren.

Kessler führt uns den ganzen technischen Apparat vor Augen und lässt dabei jeden Voyeurismus leer laufen, weil er uns zeigt, dass es – zumindest für den voyeuristischen Blick – nichts zu sehen gibt. Wir dürfen überall hindurch- und hineinlaufen, durch enge Öffnungen kriechen, auch durch die ausgesägte weibliche Genitalzone eines überdimensional vergrößerten Pornofotos, aber zum Dahinter, zum Inneren des Labyrinths werden wir nie vordringen, weil es kein Inneres gibt. Wir sind schon „innen". Die Kabel und Apparate sind wie die freiliegenden Nervenstränge des weltweiten elektronischen Gehirns, zu deren Teilen uns McLuhan längst erklärt hat.

Die Irritation und Verkehrung räumlicher Innen-Außen-Relationen kennzeichnet auch die um 1970 entstandenen Rauminstallationen von Künstlern, die sich mit dem Verhältnis von Architektur und Überwachung auseinandergesetzt haben, vor allem Bruce Nauman und Dan Graham. Man wird aufgezeichnet, sieht sich selbst von hinten oder mit zeitlicher Verzögerung an einer anderen Stelle der Installation. Nauman interessiert das Medium Video „als elektronischer Spiegel und Überwachungsanlage".[9]

In ihrem Aufbau einfach erscheinende Environments entfalten komplexe Beziehungen zwischen Sehen/Nicht-sehen und Beobachtet-werden/Nicht-sichtbar-sein. Graham nutzt bis heute einseitig spiegelndes Glas, das er erstmals 1975 in der Installation „Two Viewing Rooms" einsetzte. Zwei direkt nebeneinander liegende Raumboxen sind durch das Glas getrennt. Der eine Raum ist dunkel, der andere hell. In der hellen Box sehen die eintretenden Besucher auf einem vor dem Spiegel platzierten Videomonitor in Echtzeit ihr eigenes Bild. Im dunklen Raum können andere Besucher die Besucher im hellen Raum beobachten, die davon nichts merken – erst dann, wenn sie realisieren, dass die sie aufzeichnende Kamera sich hinter dem Spiegel befinden muss.

Auch in theoretischen Texten hat sich Graham mit der sozialen und politischen Funktion von Überwachungssystemen befasst.[10] Etliche Künstler führten das Nauman-Graham-Paradigma differenzierend weiter. Peter Weibels „Beobachtung der Beobachtung: Unsicherheit" war 1973 eine frühe europäische Variante von Naumans Closed-Circuit-Monitoranordnungen. Seit den 1980er Jahren setzt sich Julia Scher konsequent und stets unmittelbar erkennbar mit dem Thema Überwachung auseinander. Andere Künstler und Filmemacher analysieren weniger das technische Equipment als das automatisch aufgenommene Bildmaterial. Der Film „Der Riese" (1984) von Michael Klier besteht fast ausschließlich aus Bildern, die von Überwachungsanlagen in deutschen Großstädten aufgezeichnet wurden. Die „Suicide Box" (1996) des Bureau of Inverse Technology ist ein Zusammenschnitt von Bildern einer Kamera, die auf die Golden Gate Bridge in San Francisco gerichtet war und während einhundert Tagen automatisch auch die hier regelmäßig verübten Selbstmorde aufnahm.[11]

Die wichtigste theoretische Inspirationsquelle des Überwachungsthemas, des Verhältnisses von Macht- und Blickbeziehungen, waren die Schriften Michel Foucaults. In „Überwachen und Strafen" analysierte Foucault 1975 das panoptische Gefängniskonzept Jeremy Benthams aus dem frühen 19. Jahrhundert. Die Wärter konnten vom Mittelpunkt der Rundanlage aus die Gefangenen in den umliegenden Zellen überwachen, waren aber selbst für diese unsichtbar. Der niederländische Architekt Rem Koolhaas reichte 1979 einen Wettbewerbsvorschlag für den Umbau des Arnheimer Gefängnisses ein, der eine Umkehrung der einseitigen Sichtverhältnisse vorsah. Die Wärter sollten in verglasten Aufenthaltsräumen sitzen. Es versteht sich fast von selbst, dass Koolhaas' Projekt nicht realisiert wurde. Obwohl als „echte" Architektur gedacht, war es sozial so utopisch, dass es einen Status zwischen Architektur und Kunst annahm und einigen Ideen Grahams nahekam.[12]

„Viele unserer heutigen Überwachungssysteme basieren auf dem Benthamschen Prinzip – gerade wenn die Geräte nicht eingeschaltet oder ohnehin nur Attrappe sind und keinerlei Film- oder Bandmaterial oder Übertragungskabel enthalten, und trotzdem ihre Aufgabe erfüllen, erinnern sie an Benthams Panopticon",[13] schreibt Thomas Y. Levin, Kurator der Ausstellung CTRL Space 2002 im ZKM Karlsruhe. Direkt vergleichbar ist allerdings nur die Unwissenheit darüber, ob man tatsächlich überwacht wird. Die kameraüberwachten städtischen Außen- und Innenräume, die Straßen, Kauf- und Parkhäuser, Bankfilialen, Bahnhöfe oder Flughäfen sind jedoch keine „Einschließungsmilieus", wie es ein Gefängnis darstellt. Foucault hat die Einschließungsmilieus als kennzeichnend für die Disziplinargesellschaften analysiert, die vom frühen 19. Jahrhundert bis nach dem Zweiten Weltkrieg vorherrschten. Sie waren geprägt von festen Institutionen wie Gefängnis, Krankenhaus, Fabrik, Schule, Familie, die sowohl das normale soziale Leben regeln als auch die zeitweiligen oder grundsätzlichen Ausschlüsse aus diesem, etwa durch eine Gefängnisstrafe. Aber auch wer nicht ins Gefängnis kommt, durchläuft im Leben mehrere Einschließungsmilieus: Schule, Universität, langfristiger Arbeitsplatz usw.

Die Einschließungsmilieus sind zwar nicht verschwunden, aber die Disziplinargesellschaften haben sich zunehmend zu „Kontrollgesellschaften" gewandelt – ein Prozess, den Gilles Deleuze in Weiterführung von Foucaults Machttheorie zusammenfassend beschrieben hat.

In der Kontrollgesellschaft lösen sich die einschließenden Milieus als architektonische Begrenzungen auf. Statt Menschen in örtlich begrenzten Milieus Regeln zu unterwerfen und zu disziplinieren, findet Kontrolle jederzeit und überall statt, ohne dass es zu physisch direkt spürbaren Zugriffen auf den – eingesperrten – Körper kommt. Man weiß nie, wo man wirklich steht. Das Gebot lautet, sich ständig weiterzuentwickeln, neu zu lernen, den Arbeitsplatz zu wechseln – wie es der amerikanische Soziologe Richard Sennett genannt hat: ein „flexibler" Mensch zu sein. Dieser arbeitet nicht in Fabriken mit festen Hierarchien, sondern in „Unternehmen", hat es nicht mit Besitzern, sondern mit wechselnden Geschäftsführern zu tun und kann seinen Arbeitsplatz verlieren, weil die Aktionäre sich verspekuliert haben.

In seinem 1990 verfassten „Postskriptum über die Kontrollgesellschaften"[14] sieht Gilles Deleuze die elektronische Fußfessel, die eine herkömmliche Haftstrafe ersetzt, als grundlegende Metapher für die Kontrollgesellschaft. Der Verurteile muss nicht im Gefängnis verweilen, kann sich aber auch nicht wirklich frei bewegen, weil sein Aufenthaltsort immer überprüfbar bleibt. Der Erfinder des elektronisch überwachten Hausarrests, ein Haftrichter aus New Mexico, erklärte, dass er die Idee dazu aus einem 1979 erschienen Spiderman-Comic übernommen habe.[15]

Anders als die frühen Installationen von Graham und Nauman, in denen man in mehr oder weniger geschlossenen Räumen einer Art Versuchsanordnung ausgesetzt ist, wird man bei Jon Kessler eher zum Träger einer elektronischen Fußfessel. Man kann frei herumlaufen, aber das Kontrollsystem weiß immer, wo man ist. Und es kann vorkommen, dass wir auf einem der Monitore ähnlich überraschend ein vertrautes Gesicht aus dem Bekanntenkreis entdecken, wie wenn es unvermutet im Fernsehen auftaucht. Wir wussten nur nicht, dass der oder die Bekannte gerade auch die Ausstellung besucht.

Die Echtzeitübertragung stellt die Gewissheit her, dass der Bekannte wirklich in diesem Raum anwesend ist. Diesen Authentizitätseffekt sieht Thomas Y. Levin als wesentlichen Grund für das immer stärker zunehmende Interesse an Überwachungskameras und Echtzeitübertragungen: „Der ‚Wirklichkeitsanschein' (...), der in der Vergangenheit von der Indexikalität der Fotografie garantiert wurde, stellt sich jetzt durch die rhetorische Kraft der so genannten ‚Echtzeit' ein."[16]

Genau diesen Wirklichkeitsanschein führt Jon Kessler ad absurdum. Was wir auf den Monitoren sehen, sind Zeitungsfotos, Collagen, Spielzeugsoldaten, gestanzte und gebogene Bleche, ein mit schwarzer Farbe geschriebenes „Fuck You", die auf eine sich drehende Trommel gezeichneten Linien (die Duchamps Schokoladenmühle zitiert), an der Wand hängende Handpuppen, die Lichtbrechungen farbiger Folien – Bilder und räumliche Arrangements, die keinen Wert eines Wirklichkeitseffektes besitzen. Damit nimmt Kessler gleichsam postmoderne Bildtheorie wörtlich, die nach Peter Weibel „nicht von einer Weltbeobachtung, sondern von einer Bildbeobachtung ausgeht".[17]

Die Erkenntnis, dass auch die Kunst selbst Bild- und nicht Weltbeobachtung ist, prägt Jon Kesslers Auseinandersetzung mit dem 11. September und den Veränderungen des politischen Klimas auf angenehme Weise. Er verzichtet völlig auf die Idee, dass ein Kunstwerk mit terroristischen Akten und ihren realen Auswirkungen in Konkurrenz stünde. Dieser Gedanke leitete ja schon die unglücklichen Äußerungen Karlheinz Stockhausens, der wenige Tage nach dem 11. September in einem Pressegespräch in Hamburg vom größten Kunstwerk sprach, das es je gegeben habe.[18]

Vielleicht war seine Argumentation aber lediglich veraltet und wiederholte fast zweieinhalb Jahrhunderte später Überlegungen des britischen Publizisten Edmund Burke, der in seiner Philosophie des Erhabenen 1757 schrieb: „Man lasse an irgendeinem Tage die erhabenste und eindrucksvollste unserer Tragödien aufführen; man wähle dazu die beliebtesten Schauspieler; man spare keine Ausstattung und Dekoration; man vereinige die höchsten Anstrengungen von Dichtkunst, Malerei und Musik; und dann, wenn alle Zuschauer versammelt sind, gerade in dem Augenblicke, in dem die Gemüter vor Erwartung aufs höchste gespannt sind, lasse man bekannt werden, dass auf dem benachbarten Platz ein Staatsverbrecher von hohem Stande sogleich enthauptet würde: augenblicklich wird die Leere des Theaters die relative Schwäche der nachahmenden Künste beweisen und den Sieg der wirklichen Sympathie verkünden."[19]

Dass Kunst nicht als Konkurrenz zur Realität, sondern als ihre Reflektion verstanden werden sollte, lässt sich schon aus Kants Überlegungen zum Erhabenen ableiten. Indem Kant die Eigenschaft der Erhabenheit nicht auf physische Objekte, sondern auf das Erkenntnisvermögen des Subjekts bezieht, führt er den Ausweg aus der Sackgasse vor, in die Burkes Argumentation automatisch führt. Doch noch heute wird der künstlerischen Fiktion nicht nur von Stockhausen ihre relative Schwäche attestiert. Ein Jahr später blickte auch Damien Hirst mit Bewunderung auf die reale Kraft des terroristischen Aktes: „The artist Damien Hirst said last night (in einem Gespräch für BBC Online News, L. S.) he believed the terrorists responsible for the September 11 attacks ‚need congratulating' because they achieved something which nobody would ever have thought possible' on an artistic level. Hirst, who is no stranger to controversy, said many people would shy away from looking at the event as art but he believed the World Trade Centre attack was kind of like an artwork in its own right'."[20]

Die Frage, ob Terrorakte oder Kriege Kunstwerke seien, stellt Kessler gar nicht erst, denn seine Installation steht nicht in Konkurrenz zur unmittelbaren Realität und auch nicht zu der globalen Bildmacht der Medien. Seine ratternden, kreisenden, wackelnden oder wie ein Artilleriefeuer knatternden Apparaturen können kaum ein Versuch sein, mit der Wirkungs- und technologischen Macht der Instanzen zu konkurrieren, die Bilder vom 11. September oder vom Irak-Krieg global verbreiten.

Während der Leerlauf der motorisierten Eisenskulpturen Jean Tinguelys den Funktionalismus des zu Ende gehenden Maschinenzeitalters spielerisch lächerlich machte, lässt sich Kesslers bühnenhaftes und offensichtlich zusammengebasteltes Arrangement als groteske Karikatur sehen, welche die globale Verbreitung stereotyper Medienbilder als gigantische kinetische Skulptur parodiert.

Kesslers Skulpturen und Objekte aus den achtziger und frühen neunziger Jahren hatten die Tradition der kinetischen Maschinen weiterentwickelt. Erst in den letzten Jahren befasst sich Kessler auch intensiv mit der Videotechnik, die er hauptsächlich, wie bei „The Palace at 4 A.M.", in Form kleiner Überwachungskameras einsetzt, die er in seine Objekte und Installationen integriert.

Und er inszeniert damit auch ein intelligentes Was-wäre-wenn-Spiel. Er tut so, als ob sich die Entwicklung der Bildmedien bis heute wie geschehen abgespielt hätte – nur dass es nach wie vor keine Möglichkeit gäbe, Bilder aufzuzeichnen und zu speichern. Damit stünden wir mit einem Bein immer noch im 18. Jahrhundert, als Lessing das Problem der Bewegung am Beispiel einer Marmorskulptur diskutierte.

Es gab allerdings schon damals zahlreiche mechanische Apparaturen und Spielzeuge, die Bewegung erzeugten oder simulierten und auch schon den Trägheitseffekt des Auges ausnutzten wie später der Film. Wie reichhaltig diese vorkinematografische, noch mit der Welt der Wunderkammern verknüpfte Welt war, ist an der publizistisch gut dokumentierten Sammlung des deutschen Filmemachers Werner Nekes zu studieren, in der man etwa anamorphotische Zerrbilder und -spiegel, Guckkästen und Perspektivtheater oder Vexier- und Silhouettenbilder bewundern kann.[21]

Viele der Apparaturen in „The Palace at 4. A.M." erinnern an die einfache Machart und das oft raffinierte Spiel mit Flächenformen bei diesen Pioniertaten, die die Bilder zum Laufen, wenn auch noch nicht zur filmischen Aufzeichnung brachten. Demnach ließe sich die Installation auch so beschreiben, dass die Welt der mechanischen Bildmaschinen und vordigitalen kinetischen Skulpturen mit heutigen Überwachungskameras ausgestattet worden ist. Ganz ohne chemische oder digitale Aufzeichnung ginge es aber doch nicht. Dann könnte es die Reproduktionen, die an der Wand hängen, auf Ständer montiert sind oder mechanisch im Kreis gedreht werden, nicht geben.

Wenn wir noch eine weitere, leicht geänderte Beschreibungsebene wählen: tut Kessler nicht auch so, als ob das Camera-obscura-Modell der Wahrnehmung bis heute herrschen würde und nicht, wie Jonathan Crary in „Techniques of the Observer" aufgezeigt hat, im 19. Jahrhundert von einem neuronalen Modell des Sehvorgangs abgelöst worden wäre? Dass die Wissenschaft das Sehen nicht mehr als in den Augenraum fallendes Licht, sondern als Erregung von Nervenreizen beschrieben hat, führte langfristig auch zur Entwicklung von „Techniken, die das Sehen auf einer vom Betrachter und vom menschlichen Auge getrennten Ebene neu anordnen (...) Die meisten der historisch gesehen wichtigen Funktionen des menschlichen Auges werden durch Praktiken ersetzt, in denen Bilder (visual images) keinen Bezug mehr zur jeweiligen Position eines Betrachters in der ‚wirklichen', optisch wahrnehmbaren Welt haben. Diese „unerbittlich fortschreitende Abstraktion des Visuellen"[22] hat Paul Virilio, auf den sich Kessler ausdrücklich beruft, als ein „Sehen ohne Blick"[23] beschrieben. Diese Trennung aber lässt Kessler nicht zu: jedes Bild, das auf einem der Monitore erscheint, stellt sich gleichsam unserem Blick, der es zu identifizieren und damit auch räumlich einzuordnen vermag. Die Komplexität der Installation erschließt sich, wie bei einer Skulptur, indem man sie durch die eigene Bewegung immer genauer wahrnimmt und sich ein inneres Bild ihrer Gesamtform herstellt.

Die Rezeption der Installation ist auch eine imaginäre Zeichnung im Raum, und in diesem Sinne scheint Kessler von Alberto Giacomettis 1932 entstandener Tischskulptur „The Palace at 4 A.M." nicht nur den Titel übernommen zu haben. Giacomettis Palast ist ein hauptsächlich aus schmalen Graten bestehendes Gebilde, das wie eine materialisierte Zeichnung im Raum wirkt.

Kessler macht die räumlichen Beziehungen trotz aller Verwirrung immer nachvollziehbar. Sein skulpturales Denken leitet die Anordnung der unterschiedlichen Einzelelemente, was einen grundlegenden Unterschied zum Vorgehen vieler anderer Künstler ausmacht, die ihre Installationen und Performances als eine Art verräumlichte Malerei verstehen. Auch fehlt bei Kessler jedes performative Element, abgesehen davon, dass man als Besucher natürlich auch absichtliche Performances aufführen kann, die von anderen Besuchern irgendwo auf den Monitoren verfolgbar wären. Jeder, der durch die Installation läuft, sieht letztlich etwas anderes. Das bühnenhafte Arrangement Kesslers weist Parallelen zu den „Animatographen" des Regisseurs und Aktionskünstlers Christoph Schlingensief auf. Hierbei handelt es sich um Drehbühnen aus zwei sich gegenläufig zueinander bewegenden Scheiben, auf denen sich, wie bei der Installation „Kaprow City" im September 2006 in der Berliner Volksbühne, zahlreiche einzelne Räume unterschiedlicher Größe befinden. In parallelen Handlungssträngen agieren hier die Darsteller eines live produzierten und geschnittenen Films. Ein Teil der Besucher befindet sich auf der Bühne, ein anderer hinter dem Vorhang außerhalb von ihr. Die meisten sitzen im Zuschauerraum und sind fast die ganze Zeit durch eine Leinwand vom Bühnengeschehen getrennt. Sie sehen aber Direktübertragungen einzelner Szenen wie in schlechter Webcam-Qualität.

Das verworrene Spektakel um die von Jenny Elvers-Elbertzhagen gespielte Lady Di, eine Hommage an den Erfinder des Happenings Allan Kaprow, führt die unterschiedlichen Betrachtungsebenen des Publikums jedoch nicht zusammen, sondern eher in einen entropischen Verfall hinein, der auch die – zumindest bei der Aufführung, die ich gesehen habe – zunehmend stagnierende und in Einzelmomente zerfallenden Live-Handlung betrifft. Und der Einsatz der Kamera entspricht dem pornografischen Charakter der Videoüberwachung, die schon Warhol prognostizierte: die Leute würden Videokameras hauptsächlich dazu nutzen, Pornos zu drehen und ihre Nachbarn zu bespitzeln.

Auch Kessler hat das aggressiv Pornografische der universellen medialen Sichtbarkeit im Visier, aber die Art und Weise, wie er Dinge sichtbar macht, entspricht eher der Nonchalance, mit der Mitarbeiter der Universität Cambridge 1993 die erste Webcam-Übertragung im Internet realisierten. Zu sehen war die Kaffeemaschine im Computerlabor, und gesendet wurde bis 2001. Ob da jemand die ganze Zeit hingeschaut hat?

Wer bei Kessler die ganze Zeit hinschaut, wird immer wieder etwas Neues entdecken und am Ende dann doch realisieren, dass alles bleibt, wie es ist. „The Palace at 4 A.M." ist eine der raffiniertesten Medien- und Videoskulpturen, die bis heute realisiert worden sind. Man könnte in ihr, wie in den Videoskulpturen Nam June Paiks, eine Allegorie auf unser Medienzeitalter sehen. Kessler verzichtet jedoch auf direkte gegenständliche und symbolische Assoziationen wie roboterartige Figuren, Kreuze, Sterne oder Tore, wie Paik sie aus Dutzenden von Monitoren bildete. Kessler setzt weder auf die schnell erfassbare Bildhaftigkeit einer solchen „erzählerisch-gegenständlichen Gesamtstruktur“[24], noch nutzt er die physische Schwere skulpturaler Materialien, um technische Geräte mit ihrer Zerstörbarkeit an ihr symbolisch scheitern zu lassen, wie es etwa Wolf Vostell mit seinen einbetonierten Autos und Fernsehern getan hat. Statt in die Einmauerung einer „Depression Endogène“, wie Vostell eine Fernsehskulptur von 1984 nannte,[25] führt uns Kessler in ein offen erscheinendes System, von dem wir uns – wie Edgar Allan Poes Seemann – erst selbst ein Bild machen müssen, um uns orientieren zu können.

1 Marshall McLuhan, Die mechanische Braut. Volkskultur des industriellen Menschen (1951), Amsterdam 1996, S. 7.

2 Zit. n. Thomas Y. Levin, Die Rhetorik der Überwachung: Angst vor Beobachtung in den zeitgenössischen Medien, in: 7 Hügel. Bilder und Zeichen des 21. Jahrhunderts, Kat. Berliner Festspiele, Berlin 2000, Band 6: Zivilisation, S. 49–61, hier S. 49.

3 Peter Weibel, Zur Politik der Visibilität im panoptischen Zeitalter. Von Zero Tolerance zu Ground Zero, (2002), in: ders., Gamma und Amplitude. Medien- und kunsttheoretische Schriften, hg. von Rolf Sachsse, Berlin 2004, S. 407–428, hier S. 407.

4 Marshall McLuhan, Die magischen Kanäle (Understanding Media), Düsseldorf/Wien 1968, S. 50 ff.

5 McLuhan, Die mechanische Braut, a. a. O., S. 7.

6 Juliane Rebentisch. Ästhetik der Installation, Frankfurt/M. 2003.

7 „The Palace at 4 A.M.“ basiert zum Teil auf der Installation „Global Village Idiot“ 2004 bei Deitch Projects in New York. Der auf McLuhans Begriff des globalen Dorfes Bezug nehmende Titel ist auch die Überschrift einer Sammlung von *Guardian*-Kolumnen des englischen Bestsellerautors John O'Farrell mit kritischen und satirischen politischen Kommentaren unter anderem zum Irak-Krieg.

8 Die weitreichendsten Analysen von Bijls Werk sind: Frank van de Veire, Die Begräbnisszene des Gegenstandes. Zu den Installationen von Guillaume Bijl, in: Guillaume Bijl, Kat. Museum van Hedendaagse Kunst Antwerpen, Kunstverein Hannover, Freiburger Kunstverein u. a. 1996, S.17–39, und das Kapitel über Bijl in: Stefan Römer, Künstlerische Strategien des Fake. Kritik von Original und Fälschung, Köln 2001, S. 215–239. Römers Konzentration auf die konzeptuelle Seite des Werkes und Bijls Rezeption in der Kunstkritik unterbelichten allerdings den strategischen Illusionismus, der sich auch darin äußert, dass Bijl die verwendeten realen Gegenstände als Skulpturen versteht.

9 Barbara Engelbach in: Bruce Nauman: Versuchsanordnungen. Werke 1965–1994, Kat. Hamburger Kunsthalle 1998, S. 37.

10 Siehe „Video in Beziehung zur Architektur“ und Grahams Texte zu einzelnen Überwachungsinstallationen in: Dan Graham, Ausgewählte Schriften, hg. von Ulrich Wilmes, Stuttgart 1994.

11 Den besten Überblick zur künstlerischen Auseindersetzung mit dem Thema Überwachung bietet: CTRL Space. Rhetorics of Surveillance from Bentham to Big Brother, Kat. ZKM Karlsruhe, Karlsruhe/ Cambridge, Mass. 2002.

12 Rainer Metzger diskutiert Koolhaas' Projekt im Zusammenhang mit Dan Graham in seinem Buch: Kunst in der Postmoderne – Dan Graham, Köln 1996, S. 149.

13 Levin, a. a. O., S. 61.

14 In: Gilles Deleuze, Unterhandlungen 1972–1990, Frankfurt/M. 1993, S. 254–262.

15 Vgl. Levin, a. a. O., S. 54, leider ohne Quellenangabe.

16 Levin, a. a. O., S. 56.

17 Weibel, a. a. O., S. 411.

18 Auf die Frage, ob es nicht doch einen Unterschied zwischen Kunst und Verbrechen gebe, räumte er zwar ein, dass er die Terroristen schon als Verbrecher sehe, „weil die Menschen nicht einverstanden waren. Sie sind nicht in das Konzert gekommen“, doch konnte die breite Öffentlichkeit seine Haltung damals nur als unangemessenen Zynismus wahrnehmen.
Komplettes Transkript des Pressegesprächs am 16. September 1991 in Hamburg in: MusikTexte 91, November 1991, S. 69–77 (auch: http://stockhausen.org/hamburg.pdf), Zitat S. 77.

19 Edmund Burke, Vom Erhabenen und Schönen, übers. v. Friedrich Bassenge, Hamburg 1980, S. 81 f.

20 The Guardian, 11.09.2002, http://www.guardian.co.uk/september11/oneyearon/story/0,12361,790059,00.html

21 Ich sehe was, was du nicht siehst! Sehmaschinen und Bilderwelten. Die Sammlung Werner Nekes, Kat. Museum Ludwig, Köln, Göttingen 2002.

22 Jonathan Crary, Techniken des Betrachters. Sehen und Moderne im 19. Jahrhundert (1990), Dresden/Basel 1996, alle Zitate S. 12.

23 Z. B. in: Die Sehmaschine, Berlin 1989.

24 Konstanze Thümmel in einem Katalogbeitrag zu Paiks „Passage“ von 1986 in: Kunst der Gegenwart, Eröffnungskatalog des Museums für Neue Kunst/ZKM Karlsruhe, München/New York 1997, S. 202.

25 Dazu schreibt Heinrich Klotz: „Vostells *Depression Endogène* ist eine Krankheit, die zwar Bilder bietet, doch sie beinahe im Beton ersticken lässt, oder sie – schlimmer noch – nett verniedlicht.“ (Ebd., S. 276)

Jon Kessler Interview

with General Arthur Prinzhorn & Ethan Prinzhorn

Dieses Interview wurde am 5. Juni 2006 in Yonkers im Bundesstaat New York geführt. Arthur Prinzhorn ist ein hoch dekorierter Viersternegeneral, der seinen Dienst in der US-Armee im Jahr 2003 beendete. Sein jüngster Sohn Ethan wird im Jahr 2007 seinen Magister-Titel im Fach Kuratieren vom Bard College erhalten. In Yonkers wohnten die Prinzhorns Tür an Tür mit den Kesslers.

General: Kommen wir gleich zum Wesentlichen. Ist Ihre Ausstellung eine Kritik am gegenwärtigen Krieg im Irak?

Jon: Die Ausstellung ist sicherlich eine Reaktion auf unsere aktuelle Lage. Ich persönlich kritisiere zutiefst die Missachtung, die die Bush-Administration unserer Verfassung gegenüber an den Tag legt und die Art, in der das Weiße Haus vor den rechten Interessengruppen kuscht. Mit Unterstützung williger Medien hat uns unsere Regierung zum Irakkrieg verleitet, und es liegt auf der Hand, dass sich momentan dasselbe mit dem Iran abspielt. Die Ausstellung ist aber weniger eine Polemik gegen Bush als mein persönlicher Versuch, mit der Wut und der Entfremdung fertig zu werden, die ich seit einiger Zeit empfinde. Wie sehen Sie das, General? Sie haben ja in der letzten Zeit eine sehr deutliche Meinung zum Irakkrieg und zu einigen Mitgliedern der Bush-Regierung vertreten, besonders zu Donald Rumsfeld.

General: Als General trägt man eine Verantwortung für die Männer und Frauen, die unter Ihnen dienen, genauso wie Sie Verantwortung für die Zivilisten tragen, zu deren Schutz sie sich mit einem Schwur verpflichtet haben. Ich habe mich gegen Rumsfeld ausgesprochen, weil er eine vollständige Missachtung der Männer und Frauen der Streitkräfte an den Tag gelegt hat und es weiterhin tut, dasselbe gilt für seine Geringschätzung des amerikanischen Volkes. Die Weise, in der er diesen Krieg führen möchte, ist untragbar. Wir tun nichts anderes, als für die untaugliche Vision eines „neuen" Mittleren Ostens unschuldige Menschen zu opfern und als Kollateralschäden abzutun. Ich unterstütze die Verbreitung der Demokratie in dieser Region und halte es für notwendig, dass dort ein politischer Wechsel stattfindet, aber die Bush-Administration geht das Problem auf falsche Weise an. Es ist sehr wohl möglich, dass sie uns mit ihrem Vorgehen in einen Atomkrieg mit dem Iran hineinzieht, in dem es keine Gewinner, sondern nur Verlierer geben kann. Der 11. September hätte eine Warnung für Amerika sein müssen, dass es seine Augen öffnet und sieht, wie unsere Außenpolitik und die fragwürdigen Versuche, unsere Freiheit zu verteidigen, im Rest der Welt wahrgenommen werden. Aber abgesehen von einer kurzen Phase des Flaggenschwenkens haben wir einfach weitergeschlafen. Während ich das hier sage, beginnt der Kongress eine dreiwöchige Debatte über die gleichgeschlechtliche Ehe – mein Gott!

Ethan: Aber Dad, warum hast du uns dann so kritisiert, als wir gegen den Krieg auf die Straße gegangen sind? War das nicht genau ein Ausdruck der Demokratie, die Du im Mittleren Osten verbreiten möchtest?

General: Ich habe niemals euer Recht auf Protest kritisiert; ich wusste lediglich, dass er sinnlos war. Ich habe noch nie eine Regierung erlebt, die trotz der weltweiten Widerstände derart entschlossen war, in den Krieg zu ziehen.

Ethan: Dann sind wir uns wohl alle darin einig, dass wir uns belogen fühlen. – Jon, wie reagierst du mit deiner Ausstellung auf dieses Unrecht?

Jon: Wenn man mir vor fünf Jahren gesagt hätte, dass ich über dieses Thema arbeiten würde, hätte ich es nicht geglaubt. Ich fühle mich wie Michael Douglas in *Falling Down* oder wie Peter Fonda in *The Wrong Man* – man kann nur bis zu einem gewissen Grad herumgestoßen werden, und dann muss man selber handeln. 2001 habe ich Mobiles angefertigt, in die Körperabgüsse von meiner Frau, meiner Tochter und mir integriert waren. Nach 9/11 bekam ich den Anblick nicht mehr aus dem Kopf, den die Terroristen aus den Cockpits gehabt haben müssen. Deshalb entschloss ich mich, ihn nachzustellen. Dieser Impuls führte dann zu *One Hour Photo* und seitdem hat sich meine Arbeit in diese stärker politisierte Richtung entwickelt.

Ethan: Das heißt, 9/11 hat deine Arbeit deutlich verändert?

Jon: 9/11 war Anlass für die Veränderung meiner Arbeitsweise; wesentlich war aber auch, dass ich Videos in meine Skulpturen einbezogen habe. 2004, als ich die Ausstellung „Global Village Idiot" machte, hatte ich meine Arbeiten seit zehn Jahren nicht mehr gezeigt. Ich hatte nichts mehr zu sagen. Die Einbeziehung von Videos in meine Arbeit im Jahr 2001 war der Durchbruch. Man hat dies als Rückkehr zu einer lockereren, eher unfertigen Arbeitsweise verstanden. Während der 90er war meine Arbeit sehr auf die Produktion konzentriert und ich habe zahllose Stunden in meinem Studio damit verbracht, Dinge zu tun, die ich eigentlich gar nicht mag, wie zum Beispiel Aluminium polieren. Als ich dann in den jüngeren Arbeiten Maschinen mit Videomonitoren zusammengebracht habe, konnte ich die Mechanismen und Funktionsweisen der Apparate unmittelbarer und direkter darstellen. Sobald sie den gewünschten Effekt für die Kamera hervorbringen, lasse ich sie, wie sie sind.

Ethan: In den 80ern wurden Sie besonders für Ihre Skulpturen bekannt, in die Kitschelemente und gefundene Objekte integriert waren – inwiefern würden Sie diese aktuelle Ausstellung als eine Erweiterung Ihrer damaligen Ideen sehen?

Jon: Ich war immer daran interessiert, die Leute einen Blick hinter den Vorhang werfen zu lassen. Sie dazu zu bringen, aktive Betrachter zu werden, den Mechanismus dahinter zu untersuchen, ihr Vertrauen zu gewinnen und, zu guter Letzt, sich mit meinen Objekten anzufreunden, auch wenn viele davon ihre Ursprünge im Kitsch hatten. Meine ganz frühen Arbeiten spielten mit dem Bildraum, indem sie eine Dualität zwischen dem Mechanismus und den Bildschirmen herstellten, die hoffentlich mehr als die Summe ihrer Teile war. Meine neueren Videoarbeiten stellen auf vielfältige Weise eine Rückkehr zu dieser Dualität dar. Es ist lustig, dass du sagst, ich sei in den 80ern bekannt geworden. Die Veränderungen in meiner Arbeit wurden nämlich dadurch erleichtert, dass ich mich vollkommen frei fühlte, mich neu zu erfinden. Ich hatte alle meine Galerien entweder verloren oder war selber gegangen, und meine Arbeiten stießen auf wenig Interesse. Als *Artforum* dann seine Doppelausgabe über die 80er herausbrachte und ich nirgends erwähnt wurde, konnte ich ganz sicher sein, dass mich niemand mehr beobachtet.

Ethan: Ich würde auch sagen, dass es ein verbindendes Thema zwischen deinen frühen Arbeiten und deiner aktuellen Ausstellung gibt, und zwar den Fetisch. Die frühen Arbeiten scheinen mit der Idee des Fetischs in

Bezug auf Objekte zu spielen, zum Beispiel die komplexe Macht, die wir dem Kitsch beimessen. Auf ähnliche Art verweist du in deiner aktuellen Ausstellung auf die Macht, mit der wir das Ereignis – den Fetisch – 9/11 aufladen. Stimmst du dem zu? Hast du jemals Angst, in die Falle der Fetischisierung von 9/11 zu tappen, wie es so vielen passiert ist – Politikern und Künstlern?

General: Ich gehe noch einen Schritt weiter als mein Sohn und behaupte, Ihre Ausstellung hat die Ereignisse des 11. Septembers ausgebeutet!

Jon: Zunächst zu Ethans Punkt: Meine Arbeiten aus den 80ern haben sich mit dem Warenfetisch auseinandergesetzt, obwohl das niemals derartig im Vordergrund stand wie zum Beispiel bei meinen Zeitgenossen Jeff [Koons] oder Haim [Steinbach]. In den asiatisch inspirierten Arbeiten ging es eher um einen Versuch, Kultur zu fetischisieren, Exotismus und Andersheit in Ware zu verwandeln. Was die Auswirkungen der Ereignisse des 11. Septembers angeht, kann man den schädlichen Effekt gar nicht überschätzen, den Al Quaidas mörderische Inszenierung auf die Bildproduktion in unserer Kultur gehabt hat. Der 11. September ist ein Zeichen für die Verletzbarkeit der Vereinigten Staaten und der Beweis dafür, dass wir nicht länger ein Monopol auf Gewalt haben.

General: Ich weiß nicht, ob das Ihre Absicht war, aber die Geräusche, die die Apparate in Ihrer Ausstellung produzieren, hören sich wie Artilleriefeuer an. Ich habe mich gefühlt, als wäre ich zurück im Schützengraben.

Jon: Während des Sommers hatte ich nur die *Sniper*-Apparate in meinem Studio laufen. „*Sniper*" sind Mechanismen, die Kameras in Vierteldrehungen bewegen, wobei sie sich eines Maltesergetriebes bedienen. Ehrlich gesagt, hatte ich den Rhythmus des Klangs nicht vorhergesehen, den sieben dieser Maschinen erzeugen, wenn sie gleichzeitig laufen. Ich fand, dass es sich wie eine Reihe aneinander geketteter Sträflinge anhört, aber viele wurden wie Sie an den Klang von Artilleriefeuer erinnert – sichern und laden. Ich fand die Weite des Raums im P.S.1 einschüchternd und beschloss deshalb, den Raum nicht mit physischer Masse, sondern mit Blickachsen, Zielen, Sichtzonen zu strukturieren. Die *Sniper* haben auch die Verbindung zwischen Überwachung und dem neuen Auge des totalitären Krieges hergestellt, den *smart bombs*.

Ethan: Einer der vielen Widersprüche deiner Installation war der Gegensatz zwischen der Präzision, mit der die Apparate den mediatisierten „Raum" kontrollierten, und dem lockeren Aufbau der Maschinen mit den frei hängenden Kabeln.

Jon: Die Kontrolle über den mediatisierten Raum ist einer der Subtexte der Ausstellung. Die Ausstellung ist gleichzeitig Kommentar und Beispiel hierfür. Sie treibt das Prinzip der Kontrolle weiter und unterminiert es zugleich.

Ethan: 1954 stieß Michel Carrouges auf eine Strukturähnlichkeit zwischen Duchamps *Großem Glass* und Kafkas Kurzgeschichte *In der Strafkolonie.* Seiner Ansicht nach handelt es sich in beiden Fällen um geschlossene Systeme, in denen sich ein Element einem anderen aufzwingt. Das erinnert mich in mehrerer Hinsicht an deine Installation und an deine Erwähnung von Abu Ghraib.

Jon: Die Installation betrifft tatsächlich geschlossene Kreisläufe, die aufeinander ausstrahlen, ineinander greifen und sich überlagern. Auf Abu Ghraib bezieht sich der *GI-Joe* in *Theater of Ideas*, der mit heruntergezogenen Hosen und einer Narrenkappe auf dem Kopf auf einer Box steht. Um ihn herum hängen Spielzeugsoldaten vom Gerüst der Skulptur herab; sie sollen an die verbrannten und an einer Brücke aufgeknüpften US-Zivilisten erinnern. Diese wiederum sind von Bildern des Kabuki (einer hochstilisierten japanischen Theaterform) umgeben, die Furcht einflößen, Furcht, wie sie einem Soldaten aus Kansas auf Nachtpatrouille widerfahren könnte. Die roten Spritzer und Einschüsse sollen Ordnung, Kontrolle und Sinnlichkeit der Gewalt eingrenzen und verorten, ganz ähnlich dem Einsatz von Blut in Kurosawas Filmen.

General: Gewalt hat in der Tat einen sinnlichen Aspekt. Für viele Soldaten wird der Kampf neben der Pflicht auch eine Sucht. Es ist seltsam, aber als ich durch Ihre Ausstellung gegangen bin, wusste ich häufig nicht, ob ich Ihre Arbeiten mochte, weil die Bilder anziehen oder abstoßen. Gibt es eine Tradition künstlerischer Kriegskommentare?

Jon: Für mich ist Dada das beste Beispiel – die Dadaisten antworteten auf das durch die Maschinerie der Waffen verursachte Blutbad des Ersten Weltkriegs. Aber der Titel „The Palace at 4 A.M." stammt von einem großartigen surrealistischen Werk ...

General: Worauf bezieht sich der Titel?

Jon: Er bezieht sich auf eine Skulptur von Giacometti im MoMA, die denselben Namen trägt. Ich habe diese Arbeit immer wegen ihres Sinns für die Inszenierung geliebt – eine theatralische *mise-en-scène*, die den Eindruck eines Bühnenmodells für eine Traumsequenz erweckt. Als ich die Berichte unserer Truppen las, die in Saddams Palast stationiert waren, und als ich von Abu Ghraib hörte, fand ich, dass „The Palace at 4 A.M." ein angebrachter Titel für die gesamte Ausstellung wäre. Er spielt auf die traumatisch-krankhaften Zustände an, die sich zu dieser Stunde abspielen, wenn keiner zuschaut.

General: Hat sich die Ausstellung sehr verändert, als Sie sie in Hamburg aufgebaut haben? Sie schien sich so spezifisch auf den Raum im P.S.1 zu beziehen. Was ist anders an der Hamburger Ausstellung?

Jon: Das Interessante an der Ausstellung ist ihre Flexibilität. Weil die *Sniper* mit Zoomlinsen arbeiten, ist es egal, ob das Ziel einen oder 30 Meter entfernt ist, und die Ausstellung kann sich entsprechend ausdehnen oder verkleinern. In New York war sie in einem gigantischen Raum aufgebaut, der metaphorisch für den Ballsaal des Palastes stand. Die sich anschließenden kleineren Räume waren private Gemächer, in denen ich viele der figurativen Arbeiten platzierte. In Hamburg habe ich die Plakatwände benutzt, um eine Folge von Räumen herzustellen, die es dem Besucher erlaubt, sich der Gedankenwelt der Ausstellung langsamer zu nähern. Wenn man jemanden auf einem Monitor entdeckt, kann man sich nicht einfach umdrehen, um die Person zu sehen, weil diese sich höchstwahrscheinlich in einem anderen Raum aufhält. Ich habe Löcher in die Wände und Spiegel geschnitten und die Plakatwände sind so geformt worden, dass eine labyrinthischere Reise entsteht. Abschnitte der Ausstellung werden mit Taschenlampen beleuchtet, die auf den Kameras angebracht sind und wie Suchscheinwerfer funktionieren. Die Arbeit *Shock and Awe*, in der ein Foto von der Bombardierung Bagdads so auf ein Fenster geklebt ist, dass die Außenwelt zur Kulisse wird, hat in der Hamburger Installation eine skurrile und unerwartete Bedeutung gewonnen. Eine auf die Straße gerichtete Kamera filmt die Gegend, in der Mohammed Atta wohnte, als er die Zerstörung der *Twin Towers* plante. Die Idee der Ausstellung war eine Reaktion auf den Einsturz des *World Trade Centers*. Durch einen unheimlichen und bizarren Zufall kehrt die Installation damit in Hamburg an den Ort zurück, wo der 11. September geplant wurde. Es gibt auch einige neue Arbeiten für Deutschland, zum Beispiel den Rahmen eines Flugzeugfensters von meinem Hinflug nach Hamburg.

General: Sie meinen das Bild eines Fensters aus Ihrem Flugzeug?

Jon: Nein, nein ... das echte Fenster.

General: Jon, Sie haben ein Flugzeugfenster gestohlen?!

Jon: Na ja, das Flugzeug ist einfach auseinander gefallen. Das Fenster hatte sich gelockert, als wir gelandet waren, und da hab ich es in meine Tasche gestopft. In der Ausstellung ist es jetzt in ein Außenfenster der Ausstellungsräume eingesetzt, durch das man auf eine Backsteinmauer des gegenüber liegenden Gebäudes sieht. Es gibt weiter eine neue Arbeit im Eingangsbereich, in der eine Kamera in eine Silikon-Vagina eindringt. Diese Arbeit bedeutet ein doppeltes Eindringen, einmal mit der Kamera, die gleichsam telegenetisch die mediatisierte Welt der Vagina penetriert, und dann eine Öffnung in einer Plakatwand mit der Abbildung des Schambereichs einer Frau, durch die man hindurchgehen muss, um die Ausstellung zu betreten. Damit wird eine symbiotische Beziehung zwischen Kamera und Betrachter hergestellt, die sich in den Räumen als Wechselbeziehung zwischen Zuschauern, Darstellern, Voyeuristen und Exhibitionisten fortsetzt.

General: Nun, ich kann nicht behaupten, dass meine Frau gerne durch die Vagina gelaufen ist, um in Ihre Ausstellung zu gelangen. Das schien ein bisschen aufdringlich und ihr zufolge war das ein gestalterischer Fehlgriff.

Jon: Ich habe an die Beziehung von Pornografie und Krieg gedacht, die mit der Irakinvasion entstand, als Amateur-Porno-Webseiten der einzige Ort waren, an dem man unzensierte Aufnahmen der Kämpfe sehen konnte. Diese Webseiten begannen als Orte, an die man Nacktfotos von Freundinnen oder Ehefrauen senden konnte, haben sich dann aber zu Plätzen entwickelt, an denen die US-Truppen die Videos schrecklicher Kampfszenen ins Netz stellen konnten, damit ihre Familien und Freunde sich einen wirklichkeitsgetreueren Eindruck von den Geschehnissen im Irak machen konnten. Der Betrachter betritt die Ausstellung durch eine Vagina, die zu einem wandgroßen Foto gehört, das eine nackte Frau von hinten zeigt. Das Bild kommt aus einem nicht jugendfreien Magazin, aber alle expliziten sexuellen Details sind entfernt worden. Ich wollte, dass der Zuschauer in meine Ausstellung hineingeboren wird, dabei aber vermeiden, dass es sich um eine saubere, simple Geburt handelt, sondern eine, die durch das Verlangen der Medien nach Katastrophen und Nervenkitzel kompliziert wird.

Ethan: Stimmt. In der ganzen Ausstellung verwendest du Bilder, die Nervenkitzel und Gewalt widerspiegeln, deswegen macht man auch ständig diese Doppelerfahrung, indem man das Chaos einerseits interaktiv, andererseits durch das Standbild erfährt. Könntest du etwas mehr über deine Verwendung von angeeignetem Bildmaterial und Originalfotos sagen? Wie verhält es sich zum Beispiel mit der Arbeit *The Office*?

Jon: Ich habe mit vier Coffee-Table-Büchern begonnen, die ich in der *Strand*-Buchhandlung in New York gekauft habe. Alle handelten von Macht und Gewalt im heutigen Amerika. Die Bücher waren *Ready-mades* und wurden Ausgangspunkt meiner Erzählungen. Ich habe die Seiten herausgerissen, zerschnitten, bemalt und collagiert, um neue Geschichten zu formen. Das hat meine Arbeit in Schwung gebracht. In New York und Hamburg ist *The Office* das erste Stück, das man sieht. In ihm geht es um das Eingelulltwerden am Arbeitsplatz, eine mit Antidepressiva versetzte Monotonie, die unaufhörliche Ruhe, nur ein Vorspiel zum Sturm. Ich stelle mir *The Office* auch als ein Rekrutierungsbüro vor, um längst abgewertetes menschliches Material ohne Bildung für den Krieg anzuwerben, in der Terminologie Cheneys Kollateralschäden, wenn später etwas zustößt.

Ethan: Und welche Rolle spielen die *Swans*?

Jon: Die *Swans* verweisen auf die Verschönerungs-Reality-Shows wie *The Swan* oder *Nip/Tuck*. Ich habe Abbildungen aus populären Frauenzeitschriften herausgeschnitten und neu angeordnet. Diese Methode geht auf die Fotokollagen der Dada-Künstler, etwa Hannah Höch und John Heartfield zurück. Die *Swans* folgen eher einer Linie, an deren Anfang Picassos kubistische Blech-Gitarre steht. Die Betrachter vervollständigen die an der Wand angebrachten Darstellungen der Frauen, wenn sie sich ihnen nähern; eine Überwachungskamera fängt die Situation auf dem Schirm ein. Die *Swans* bewegen sich in einer uns beherrschenden Welt der Mode, der Warenfetischisten und Celebrity-Kultur. Aber sie erobern etwas Macht zurück, indem sie anstelle der Bilder, die uns kontrollieren, eigene Bilder produzieren.

Ethan: Jon, ich finde es leicht beunruhigend, dass alle Bilder, die du auseinander gerissen hast, Frauen zeigen. Gibt es dafür einen Grund?

Jon: Ich habe an anderen Stellen der Ausstellung die Gesichter von Männern zerfetzt, aber aus anderen Gründen. Die *Swans* sind alle Frauen, weil wir in einer Kultur leben, in der der weibliche Körper zum Objekt wird, und angesichts von Fernsehsendungen wie *The Swan* oder *Nip/Tuck* stellen diese Arbeiten eine Art Erweiterung dessen dar, was sowieso schon passiert. Die Gesichter und Körper von Frauen werden auf eine Weise verzerrt, dass das Künstliche zum Ideal wird. Individuelle Züge werden durch Formen und Texturen ersetzt, die man niemals in der Natur findet; ungerade Nasen, schiefe Zähne, dünnes Haar, kleine Brüste werden ausradiert; und echte Menschen mutieren zu bizarren Fälschungen. Lippen und Brüste und Brauen werden unglaublich deformiert, sind aber so allgegenwärtig geworden, dass sie anfangen, fast schon normal und natürlich auszusehen. Es hat mir immer Spaß gemacht, damit in meiner Arbeit zu spielen – mit künstlicher Natur.

General: Wer ist der haarige Typ? Der ist doch auf jeden Fall ein Deserteur.

Jon: Ja, General, das ist er. Der wurde ursprünglich für *Party Crasher* gemacht. Ich habe eine englische Spielzeugfigur verwendet und sie mit Pferdehaaren versehen. Tatsächlich stellt er einen Araber dar. Instinktiv wollte ich zuerst ein Selbstportrait herzustellen: einen alternden, sonnengebräunten Hippie: ich, wenn ich ein paar andere Lebensentscheidungen getroffen hätte. Aber als ich ihn dann fertig im Studio hatte, fragten mich Leute, die vorbeikamen, wer der obdachlose Typ sei, oder der Terrorist. So wurde er der Hippie/Obdachloser/Terroristen-Typ, eine Art flottierender Signifikant. In „The Palace" kehrte er als Hauptfigur von *Evolution* wieder. Indem ich zwei unterschiedliche arabische Köpfe verwendete, habe ich aus dem Ready-made vier Charaktere geschaffen. Im Übergang erwacht er in der glimmenden Glut des World Trade Centers, küsst auf einer Blog-Seite die Truppen im Irak, wird zum Viersternegeneral – und schließlich – nachdem er sich einer Botox-Behandlung unterzogen hat – zu einem republikanischen Senator.

General: Möchten Sie damit sagen, dass diese Typen ein und derselbe sind?

Jon: Sie stehen in diesem Augenblick alle auf derselben Weltbühne, sie reiben sich aneinander und reagieren aufeinander.

Ethan: Könntest du etwas über die Arbeit sagen, in der du Bildaufnahmen von Google Earth verwendet hast, in der gezeigt wird, wie das P.S.1 und Phoenix in die Luft gejagt werden?

Jon: Wie bei den Straßenkameras wollte ich den Aufbau der Ausstellung weiter verkomplizieren, indem ich Bildmaterial hinzufügte, das ich selbst in meiner eigenen Low-Tech-Weise mit Kameras, Fotos und Apparaten hätte aufnehmen können. Nebenbei ist Google Earth die erstaunlichste Überwachungsquelle, derer sich Künstler bedienen können.

Ethan: Würdest du deine Arbeit als interaktive Kunst bezeichnen?

Jon: Schon seit vielen Jahren bezieht meine Arbeit die Betrachter ein, und in diesem Sinn ist sie interaktiv. 1985 bei *Isolated Masses* gab es einen Strahler, der den Betrachter langsam wärmte, sobald er sich der Arbeit näherte. *Path of a Carp* von 1987 hatte ein elektrisches Auge und eine Stimmkarte, die den Betrachter auf Japanisch begrüßte. In meiner neuen Arbeiten hat sich alles verändert, als ich den Hintergrund von *Party Crasher* entfernte und der haarige Kerl denselben Raum einnahm wie die Betrachter. Diese Prämisse, die Betrachter einzubeziehen, hat sich in *Heaven's Gate* und *Gisele and the Cinopticon* fortgesetzt und ist in „The Palace at 4 A.M." dann förmlich explodiert. Die Betrachter interagieren auf jeden Fall mit der Ausstellung, ob sie wollen oder nicht, durch ständiges In-das-Werk-Eintreten, indem die Blickachsen der Kamera vervollständigt oder unterbrochen werden.

Ethan: Wie viel von der Ausstellung hattest du vorher geplant?

Jon: Der ursprüngliche Vorschlag, den ich dem P.S.1 gemacht habe, sah überhaupt nicht wie das Endergebnis aus. Die Ideen für die Ausstellung haben sich rasend schnell weiterentwickelt, als ich begann, mit den Bildern zu arbeiten, die ich aus den Büchern gerissen hatte, und als ich während des Sommers mehr Zeit im P.S.1 verbrachte. Die Ausstellung ist, je weiter ich kam, auf eine organische Weise gewachsen und hat die Richtung gewechselt, auf ganz ähnliche Art, wie sich meine einzelnen

Skulpturen entwickeln. Am Ende stellte sich „The Palace at 4 A.M." dann als viel stärker auf Bildern basierend heraus als meine vorherige Ausstellung „Global Village Idiot".

General: Ich erinnere mich, dass Ihr Vater eine Werkstatt im Keller Ihres Hauses hatte, aber ich kann mich nicht daran erinnern, dass Sie viel Zeit dort verbracht hätten. Haben Sie eine Ingenieursausbildung?

Ethan: Wie willst du denn wissen, ob er dort Zeit verbracht hat? Du warst doch niemals da!

Jon: Mein Vater war ein Bastler und mein Bruder ist ein Ingenieur, die Gene liegen also in der Familie. Ich hab im Laufe der Zeit einige Sachen aufgeschnappt und die Apparate basieren auf dem Trial-and-Error-Prinzip. Ich hatte die Idee zu einem Mechanismus, der unterbrochene Bewegungsabläufe vollzieht und eine Kamera um 90 Grad dreht. Ich habe Wochen lang daran gearbeitet und als ich endlich Erfolg hatte, kam jemand in mein Studio und teilte mir mit, dass es sich dabei um ein Malteserkreuzgetriebe handelt, das schon vor Jahrhunderten für Uhrwerke erfunden worden ist.

Ethan: Du sprichst häufig über das Spielerische in deiner Arbeit. Wie baust du das Spielerische in den Prozess ein?

Jon: Das ist schwierig. Die Herstellung von Arbeiten, die funktionieren sollen, beinhaltet viel methodische Mühen und Technik, deswegen muss ich Räume in meinen Werken schaffen, die eine intuitivere, schnellere Arbeitsweise zulassen. Beinahe alle meine Skulpturen beginnen damit, dass ich in meinem Studio auf dem Boden sitzend mit einer Videokamera und irgendwelchen Fundstücken herumspiele.

Ethan: Es ist klar, wie die Flugzeugbilder in deinen Installationen funktionieren. Warum hast du Ausschnitte von *Hummern* einbezogen?

Jon: Es ist mir jedes Mal zum Weinen zumute, wenn ich an den *Earth Day* 1970 denke, an dem wir in der High School ein Auto beerdigten. Wer hätte gedacht, dass 36 Jahre später die Leute in *Hummern* als Statussymbol herumfahren würden? Thomas Friedmann hat in seinem Kommentar „In My Next Life" in der New York Times dazu alles gesagt: „Dann möchte ich endlich einen Hummer besitzen – mit Aufklebern der amerikanischen Flagge über die ganze hintere Stoßstange, weil Hummer-Besitzer im Durchschnitt ein bisschen patriotischer sind als Sie und ich. Ja, ich will die Mutter aller Benzinschlucker fahren, die einen so hohen Verbrauch hat, dass man von Tankstelle zu Tankstelle fahren muss. Ja, ich will meinen Hummer fahren und niemals darüber nachdenken müssen, dass ich, weil ich so viel Öl verbrauche, Überweisungszahlungen an die schlimmsten arabischen Regime tätige, die Geld an islamische Wohltätigkeitsorganisationen überweisen, die Geld an Islamschulen überweisen, die den Kindern Intoleranz und Antipluralismus predigen und dass sie die Ungläubigen hassen sollen. Und wenn dann eines Tages einer der Absolventen dieser Islamschulen sich dem Djihad in Fallujah anschließt und den Sohn meines Nachbarn tötet, der bei den U.S. Army Rangers ist, dann möchte ich in meinem Hummer zu seiner Beerdigung fahren … und auf dem Weg zweimal tanken müssen."

Ethan: Das scheinen die Punkte zu sein, die du in deiner Ausstellung verknüpfen willst.

Jon: Ja, in der Ausstellung geht es um Verbindungen, die sich aus einer kausalen Vernetzung von Elementen ergeben. Das geschieht durch die Kombination von Kamerabewegungen mit der Schnittfunktion der Maschinen, aber auch durch die Position des Betrachters und seine kognitive Erfassung der Arbeit. Das kriegslüsterne Bestehen auf Angriff, die Antikriegsproteste, die Konsumkultur, die uns alle betäubt und verwirrt, Bush, der sich *Shock and Awe* im Fernsehen ansieht, du, der du dir Bush ansiehst, wie er zuschaut, jemand, der dir zuschaut, wie du Bush anschaust… Meine letzte Arbeit, *The Drowned World*, bezieht sich auf den Hurrikan Katrina, eine Tragödie, die klar zeigt, dass wir nicht in der Lage sind, auf die Bedürfnisse unserer Bürger angemessen zu reagieren, insbesondere wenn sie arm und schwarz sind. Unsere Regierung sorgt sich eher darum, Baumaßnahmen in Bagdad zu finanzieren, auch wenn dort kaum etwas gebaut wird und das Geld an Halliburton geht. Ein wesentlicher Anstoß für diese Ausstellung war der Film *Apocalypse Now* und Coppolas Vorbild einer narrativen Aufbereitung der Geschichte des Vietnamkriegs und seines Wahnsinns. Der Unterschied ist natürlich, dass *Apocalypse Now* nach dem Krieg gemacht wurde. Deswegen war ich so fasziniert von der Fernsehsendung *Over There*, einem Dokudrama über unsere Truppen im Irak. Es gibt kein besseres Beispiel für die Unmittelbarkeit, mit der das Leben von den Medien erfasst wird. Ich glaube, dass wir an einen Punkt gelangt sind, an dem sich Erfahrung und Bild getrennt haben. Das Bild ist erfahrungsfrei geworden, sodass wir durch die Nachrichten, das Internet und die verschiedensten anderen Medien auf ein Überangebot von Bildern treffen, das Paul Virilio als *Visual Crash* bezeichnet.

General: Überall Kameras, wohin man geht, ich habe es satt, ständig gefilmt zu werden. Vielleicht ist es ja ein Abschreckungsmittel gegen Täter oder vielleicht sammeln wir einfach visuelle Information im Übermaß, weit mehr, als wir jemals verwenden können. Wie man es auch sieht, das Militär kann vermelden, bei der Erfassung audio-visueller Informationen weit vorn zu liegen.

Ethan: Klar, unter Mithilfe von AT&T und weil Persönlichkeitsrechte unberücksichtigt bleiben…

General: Du weißt, was sie sagen, Sonnyboy: Wo gehobelt wird, fallen auch die Späne. Jon, können Sie uns Details über die Verwendung von Überwachungstechniken in Ihrer Arbeit verraten?

Jon: Überwachung als mediatisierte, durch Angst gesteigerte kulturelle Erfahrung ist ein elementarer Bestandteil meiner gesamten Arbeit seit 9/11. „The Palace at 4 A.M." und einige meiner neueren Arbeiten antworten auf den Irakkrieg und die widerrechtliche Besetzung des Landes. Anders als „Global Village Idiot", das acht oder neun singuläre Skulpturen zeigte, bezieht „The Palace at 4 A.M." die Betrachter mit ein, macht sie zu überwachten Subjekten und verweist auf die zwiespältige Rolle der Überwachung. Das Spiel von Kamera, Bild und Monitor ist überall präsent und die Betrachter schaffen als Mitspieler ständig neue Situationen. Die Rezipienten werden so zu aktiven Teilnehmern und Mitgestaltern der Installation.

Ethan: Auch wenn die Ausstellung sich stark der Überwachung und den Medien zuwendet, so insistiert sie doch auch auf unmittelbarer Erfahrung...

Jon: Die Ausstellung will auch sinnliche Erfahrungen vermitteln – durch die Wärme, die Bildschirme und technische Apparate erzeugen, den Gummigeruch, der von Unmengen elektrischer und koaxialer Drähte und Schläuche ausgeht, und nicht zuletzt das Geräusch der *Sniper*, die das bedrohliche Entsichern von Waffen hörbar werden lassen. Wenn „The Palace" auch konzeptionell die Rolle der Medien und unsere Beziehung zu ihnen in den Vordergrund stellt, so geht es doch immer um unmittelbare menschliche Erfahrungen. Die Maschinen und Apparate in meinen Installationen vermitteln ganz andere emotionale Erlebnisse, als man sie herkömmlich von Videoarbeiten in dunklen Räumen gewohnt ist.

Ethan: Und doch charakterisiert die Ausstellung die Bildproduktion unserer Zeit ...

Jon: Stimmt. Mit „The Palace at 4. A.M." will ich belegen, dass gegenwärtig die visuellen Stimuli den Betrachter übersättigen und dass die Welt nur noch Bühne eines ständigen Wechsels von Bildern des kollektiven Begehrens ist. Reality-Shows und Krieg in Bildern sind Belege für das heutige Bedürfnis nach Sensation und Bedeutung und Legitimation für die voyeuristischen Tendenzen in unserer Gesellschaft. Die Ausstellung lässt Zeit und Bilder verschmelzen und schafft komplexe Situationen, in denen der Betrachter gleichzeitig Zuschauer, Darsteller, Voyeur und Exhibitionist ist. Sollten wir mit Überwachungskameras leben müssen, dann stellt sich für mich nicht die Frage *Wie können wir die Kamera zerstören?*, sondern *Wie ist es möglich, das überwachte Bild zu unterlaufen und selbst die Regie zu übernehmen?*

Zeitreise zu den Ursprüngen kultureller Identität

Harald Falckenberg

„Deutschland hat Russland den Krieg erklärt. Nachmittags Schwimmschule"
(Tagebuchaufzeichnung Franz Kafkas vom 14. August 1914)

Jon Kesslers groß angelegte Multi-Media-Ausstellung „The Palace at 4 A.M." im P.S.1 Contemporary Art Center 2005 war ein Riesenerfolg. Als mich der Düsseldorfer Galerist Hans Mayer aus New York anrief und mir vorschlug, die Installation in meinen Harburger Fabrikräumen zu zeigen, sagte ich spontan zu. Noch kannte ich nur Details. Nach Hans Mayers Bericht sollte es um Bush und eine kritische Aufarbeitung des Irak-Krieges gehen. Erst beim Aufbau der Arbeit wurde Kessler und mir klar, worauf wir uns eingelassen hatten. Der gedankliche Ausgangspunkt der Arbeit ist die Vorstellung einer letzten Momentaufnahme des finalen Bildes, das die Terroristen unmittelbar vor dem Einschlag ins World Trade Center hatten. Die Bilder vom Aufprall der Flugzeuge in den Türmen durchziehen die Installation vom Anfang bis zum Ende.

Als Werk des 9/11 und seiner Folgen kehrt die Arbeit in Harburg an seinen topografischen Ursprungsort zurück. Nur wenige hundert Meter vom Ausstellungsort entfernt – in der Marienstraße – war das Attentat auf das WTC von Mohammed Atta und seinen Komplizen geplant worden. Als Kessler dies erfuhr, installierte er sogleich eine Überwachungskamera in Richtung Marienstraße, die alle Ortsbewegungen auf mehreren gut verteilten Monitoren der Arbeit wiedergaben. Des weiteren ließ er den Plastikrahmen eines Kabinenfensters aus einem Flugzeug – angeblich war dieser ihm während des Fluges nach Hamburg auf den Schoß gefallen – auf einem Außenfenster der Ausstellungsräume anbringen.

Innerhalb weniger Wochen war das meine zweite Begegnung im Zusammenhang mit dem 11. September in der Kunst. Ich hatte kurz zuvor einige Arbeiten von Paul Thek aus der Zeit um 1970 erworben. Thek war ein Protagonist des Widerstands gegen den Bau des WTC. In Zeichnungen platzierte er riesige schwarze Kreuze zwischen die Twin Towers als Symbol ihres Untergangs. Thek vermochte wie viele der New Yorker Mitbürger nicht einzusehen, warum ein alt eingesessenes, multikulturell gewachsenes Viertel dem Erdboden gleich gemacht wurde, um einem Zentrum des internationalen Kapitalmarktes zu weichen. Das Gemälde von 1972 (Abb. 1) zeigt einen Blick aus seinem Studio auf die Türme kurz vor ihrer Fertigstellung. Das Bild ist mit heißen Kartoffeln umrahmt. Ein Wortspiel – hot potatoes ist nicht nur ein Kinderspiel, in dem heiße Kartoffeln bis zum Abbruch der Musik weitergereicht werden und sich die Letzten ihre Hände verbrennen. Als hot potatoes werden im Börsengeschäft auch faule Wertpapiere verstanden, die, solange der Markt es hergibt, mit Profit weiter veräußert werden und schließlich auffliegen. Die Beobachtung und

Überwachung des Außenbereichs erfolgt bei Thek und Kessler wie in Hitchcocks Klassiker durch ein „Fenster zum Hof“ als Ausdruck voyeuristischer und paranoider Zwangsvorstellungen bis hin zum Verfolgungswahn.

Und ein weiterer Zusammenhang. Im April 2007 feierte das Centre Pompidou mit der Ausstellung „Air de Paris“ seinen 30. Geburtstag. Angelpunkt der Schau war der 1974 entstandene 16-Millimeter-Film „Conical Intersect“ von Gordon Matta-Clark (Abb. 2). Es geht um eine „Cutting“-Aktion des Künstlers, der im Umkreis der Situationisten zu den erbitterten Gegnern des Beaubourg-Projektes gehörte. Matta-Clark durchbrach die Außenfassaden und Innenräume von zwei Abbruchhäusern in unmittelbarer Nachbarschaft des Neubaus in einer Achse, die zentral auf die Mauern des künftigen Museums gerichtet waren. Matta-Clark hätte seine „Cuttings“, die man ihm nur an vor dem Abriss stehenden Häusern genehmigte, lieber an Gebäuden durchgeführt, die, wie er selbst sagte, „brandneu, frisch und in keiner Weise bereit für die Axt“ gewesen wären.[1]

Die Verknüpfung zeitlich und in ihrer kausalen Entstehung scheinbar weit auseinander liegender Sachverhalte zu einer aktuellen Situation ist verblüffend. Zufall? Man sollte vorsichtig mit Begriffen wie Zufall, Situation und Ereignis umgehen. Es gibt nicht wenige, die den Zufall als Norm und die Ordnung als Ausnahme ansehen.[2] Diese Aussage gewinnt an Gewicht, wenn unter Ordnung nur bewusste, logisch nachvollziehbare Zusammenhänge verstanden werden. Nach den neuesten Erkenntnissen der kognitiven Psychologie und Hirnforschung steht fest, dass der zerebrale Raum durch Umordnung und Umschrift der Erinnerungsspuren einer ständigen Veränderung unterliegt und als Umschlagplatz kodierter, verdrängter, nachträglich aufscheinender und wiederkehrender subjektiver und kollektiver Bedeutung wesentlich – Forscher sprechen von 90 Prozent aller Wahrnehmungs- und Erkenntnisvorgänge – im Unbewussten verortet ist.[3] Nach der Vorstellung von Gilles Deleuze und Félix Guattari schneidet der Mensch mit der Kunst, der Wissenschaft und der Philosophie als den drei großen Formen des Denkens Ebenen in das vom Zufallsprinzip beherrschte Chaos, um sich von ihm nicht zerreißen zu lassen.[4] Immer geht es um den Versuch, das sichere Ufer zu erreichen, verbunden mit Hoffnungen, die schon in der nächsten Flut wieder zu versinken drohen. Es ist vornehme Aufgabe des Künstlers, die Täuschung und Selbsttäuschung der Menschen im Endlichen zu durchlaufen, um diese dann umso überzeugender zu entlarven, zu ent-täuschen und frei zu sein für eine schöpferische Praxis jenseits des bloß Zeitgemäßen und Angesagten.

Das Verhältnis der Architekten zu den Künstlern ist seit jeher eine verhängnisvolle Affäre. Allein das Thema Museumsbau und Kunst füllt Bände. Die Architektur setzt die Wunschvorstellungen der Auftraggeber nach Bedeutung und Größe um; die Kunst ringt um Unabhängigkeit und Autonomie. Wenn auch angesichts der Opfer menschlich und sozial kaum vermittelbar, war es deshalb folgerichtig, wenn Künstler Karlheinz Stockhausen und Damien Hirst die Sprengung der Twin Towers als Fanal gegen den Kapitalismus feierten. Für Gordon Matta-Clark hat es Dan Graham Mitte der 80er Jahre auf die griffige Formel gebracht: „Der Architekt baut auf, der Künstler zerstört.“[5]

Gern hätten wir dazu Minuro Yamasaki (1912–1986) gehört, den aus Japan stammenden amerikanischen Architekten des WTC. Kommentare zum Untergang seines Hauptwerkes sind ihm erspart worden, gewiss gut so. Fast 30 Jahre zuvor hatte die von ihm 1950 entworfene und im Juli 1972 auf behördliche Anordnung durch Sprengung beseitigte Sozialsiedlung Pruitt-Igoe in St. Louis Geschichte in Sachen Dekonstruktion geschrieben (Abb. 3). Offiziell waren die verwahrlosten sozialen Verhältnisse Anlass des Abbruchs. Eigentlich ging es aber um die radikale Aufgabe ideologischer Vorstellungen, die sich mit der Siedlung als einer neuen Heimat für eine moderne Gesellschaft ohne Standesunterschiede verbanden. Pruitt-Igoe ist für die publizistischen Verkünder der Postmoderne Charles Jencks und Tom Wolfe Symbol für das Ende der Moderne. „Tod der Moderne“ – mit dieser knappen Feststellung eröffnete Jencks 1977 seine Abhandlung „The Language of Post-Modern Architecture“. 1981 folgte die Abrechnung von Tom Wolfe in seinem Essay „From Bauhaus to Our House“. Und auch in Deutschland zeigte Pruitt-Igoe Wirkung. Die 1980 gegründete Berliner Experimental-Band „Einstürzende Neubauten“ bezog ihren Namen auf die topoklastische Beseitigung der Siedlung.

Jon Kesslers Werk steht in der Tradition der Postmoderne. Als Spross einer etablierten jüdischen Familie aus New York sollte er das Bekleidungsgeschäft seines Vaters übernehmen. Er wollte aber Künstler oder Musiker werden und es sollte sich herausstellen, dass er beides wurde. Ihm ging es darum, die mimikryhafte Anpassung jüdischer Tradition an die amerikanische Kultur zu durchbrechen. Seine 1989 bis 1991 entstandene, bislang wenig bekannte Werkserie „Broken Buildings“ bezieht sich auf die Architektur. Kessler rechnet mit der Idee, über Hüllen, Formen und Funktionszusammenhänge Identität zu gewinnen, ab. Ihm ging es um das Skelett und die Mechanismen jenseits der Fassaden.

Insofern unterscheidet sich die Serie der „Broken Buildings“ weniger von seinem seit 1983 entwickelten Hauptwerk der kinetischen Skulpturen, als es der äußere Augenschein zunächst suggerieren mag. In diesen merkwürdigen Objekten verarbeitet Kessler Versatzstücke multikultureller Herkunft, die – als Kitsch, Nippes und Trash kommerziell ausgebeutet – nur noch Reflex lange zurückliegender geschichtlicher Zusammenhänge sind. New York, melting pot der Kulturen, ist der ideale Ort für Metamorphosen dieser Art. Kessler verweigert sich dem amerikanischen Traum nach Repräsentation, Macht und Reichtum. Er versteht Kitsch, anders als Jeff Koons, nicht als aufpolierte Jagdtrophäen einer auftrumpfenden Mittelschicht. Er widersetzt sich dem kalten Gedanken des Pop und der Appropriation Art und verweigert sich dem freien Formenspiel mit objets trouvés im Nouveau Réalisme. Sein Umgang mit den Objekten ist vom Respekt für die historischen Vorlagen geprägt und von der romantischen Idee geleitet, sie vor dem Verfall auf der Müllhalde des Begehrens zu retten. Neu kombiniert und reanimiert entstehen so eher zufällige und seltsam verfremdete Dokumente eines der Vergangenheit verpflichteten Alltags. „Ich bin kein Schachspieler“, sagt Kessler, „sondern Bildhauer und handele intuitiv und spontan. Meine Kunst entsteht anders. Ich weiß nie, wohin mich meine Kunst trägt, bis ich da bin.“[6]

Die Aura der Antiautorität und die Abrechnung mit der Welt der Väter ist sein zentrales Thema. Von Louise Bourgeois stammt der Satz: „Du musst deine Vergangenheit jeden Tag vergessen. Und sie akzeptieren. Und wenn du sie nicht akzeptieren kannst, musst du Skulpturen machen.“[7] Sie versteht die Bildhauerei als ungezügelte Lust an der Dekonstruktion und der symbolischen Zerstörung des Vaters – Ursache eines immer wieder entfachten Kindheitstraumas. Kesslers Werk der 80er Jahre lässt sich kaum anders als ödipale Serienschändung verstehen. Der Kunstbetrieb mit seinen Diskursen der Theorie bleibt Kessler fremd. Er versteht sich als Einzelgänger: „Ich habe immer eine Position abseits des Kampfgeschehens eingenommen. Das ist eine gute Position, wie ein Platz in der ersten Reihe Mitte im Theater.“[8] Kessler hat sich abgesetzt in eine fiktionale Welt der Esoterik und Kontemplation fernab der Väter mit ihren kommerziellen Verwicklungen. Mit der retrospektiv angelegten Ausstellung „Jon Kessler's Asia“ in der Kestner-Gesellschaft in Hannover und der Neuen Galerie am Landesmuseum Joanneum in Graz findet diese Phase der Emanzipation Kesslers 1994 einen vorläufigen Abschluss.

Die frühen 90er Jahre gerieten für viele Künstler zum Albtraum. Lange Zeit während des Kunstbooms der 80er Jahre umhätschelt und umgarnt, wurden sie nach dem Zusammenbruch des Marktes von den Galeristen von heute auf morgen fallen gelassen. Sie schlossen sich zu Selbsthilfegruppen zusammen und versuchten, sich als Kritiker, Kuratoren und Buchautoren durchzuschlagen. Nicht wenige gingen als Lehrer an die Kunstakademien. Dies war auch für den Außenseiter Jon Kessler der angemessene Weg. Seit 1994 unterrichtet er an der Columbia University's School of the Arts, 2000 bis 2005 als Dekan der Visual Arts Division. Im Schutz der Hochschule und inspiriert durch seine Lehrtätigkeit hatte er Gelegenheit, seine Kunst „abseits des Kampfgeschehens“ weiterzuentwickeln. Zunehmend konzentrierte er sich dabei auf die medientheoretischen Zusammenhänge seiner Arbeit.

Mit „The Palace at 4 A.M.“ ist Kessler ein Durchbruch gelungen. Die zwanghafte, „Ich“-psychologisierende Auseinandersetzung mit der eigenen Identität hat er abgeschlossen zugunsten einer freien Wahrnehmung, in der sich Beobachtungs- und Überwachungssysteme bis hin zum Rezipienten überlagern, der die Beobachter beim Beobachten beobachtet. Durch diesen Schritt wird nachvollzogen, dass der Beobachter selbst Teil der Wirklichkeit und des Systems ist, welches er beobachtet – einem Mix aus vorgefundener, objektiver und subjektiver Wirklichkeit. „The Palace at 4 A.M.“ ist eine eminent politische Arbeit. Sie ist, nicht zuletzt aus der persönlichen Sicht Kesslers, Beleg einer Entwicklung, die in der Medientheorie als kognitive Wende bekannt geworden ist. Danach wird die Wahrnehmung weniger durch das Auge als angeblich objektives Organ[9] als durch die Kognition und durch das Gehirn als Zentralorgan der Erkenntnisvorgänge gesteuert. Der Medientheoretiker Peter Weibel sieht die kognitive Wende in seiner Abhandlung zur „Wahrnehmung im technologischen Zeitalter“[10] durch zwei Grunderfahrungen gekennzeichnet: zum einen dadurch, dass die apparative, maschinengestützte Wahrnehmung die Steuerung von Scheinkörpern, -bewegungen und -räumen im Rahmen einer immer komplexeren Trompe-l'œil-Technologie ermöglicht; zum anderen dadurch, dass die Welt beobachterabhängig wird. Mit dem Paradigmenwechsel von der passiven Erfassung zur aktiven Gestaltung im Sinne der kognitiven Psychologie und Neurowissenschaften konstituiere sich die Wahrnehmung als Prinzip eigener Konstruktion.

Als Werkschaffender des Neuesten ist Jon Kessler das Älteste gelungen, die Rückkehr nach Mesopotamien, dem Zwei-Strom-Land, zu den anthropologischen und topografischen Ursprüngen unserer Sprach- und Zeichenkultur und an den Ort des Turmbaus zu Babel, einem der frühesten Monumente menschlicher Hybris (Abb. 4). Bleibt die Frage nach der jüdischen Identität. Ich könnte mir vorstellen, dass Kessler eine Antwort finden würde, wie sie Franz Kafka in seiner Tagebuchaufzeichnung vom 8. Januar 1914 formuliert hat: „Was habe ich mit Juden gemeinsam? Ich habe kaum etwas mit mir selbst gemeinsam und sollte sehr ruhig in einer Ecke stehen, zufrieden damit, dass ich atme.“

1 Gespräch mit Liza Bear, Gordon Matta-Clark: Splitting the Humphrey Street Building, in: avalanche, Dezember 1974, S. 34–37, abgedruckt in: Gordon Matta-Clark, London/New York 2003, S. 163–169, S. 167.

2 Dazu näher Henning Ritter, Frankfurter Allgemeine Zeitung, 7. April 2007, mit Nachweisen, insbesondere auch in Auseinandersetzung mit Sigmund Freuds diesbezüglicher Abhandlung über „Das Unheimliche“ (1919). Und ein weiterer Zufall? Einige Tage nach Abschluss dieses Textes am 8. Juni 2007 verstarb der Meisterdenker des Zufalls, der amerikanische Philosoph Richard Rorty. Die Aussagen seines zentralen Werkes „Contingency, Irony and Solidarity“ (1989) sind durch die heutigen Erkenntnisse der kognitiven Psychologie und der Hirnforschung bis ins Detail bestätigt worden. Vgl. Harald Falckenberg, Beitrag zum 150. Geburtstag von Sigmund Freud, in: Aus dem Maschinenraum der Kunst, Hamburg 2007, S. 60 ff.

3 Vgl. Harald Falckenberg, Fn. 2.

4 Gilles Deleuze/Félix Guattari, Was ist Philosophie?, Frankfurt am Main 1996, S. 234.

5 Dan Graham, Gordon Matta-Clark, in: Sabine Breitwieser (Hrsg.), White Cube/Black Box, Generali Foundation, Wien 1996, S. 215–237, S. 230.

6 Jon Kessler im Gespräch mit Lynne Cooke, in: Jon Kessler's Asia, Katalog zur Ausstellung der Kestner-Gesellschaft, Hannover, und der Neuen Galerie am Landesmuseum Joanneum in Graz, Hannover 1994, S. 54.

7 Zit. nach Claudia Steinberg über Louise Bourgeois, Kunstzeitung, Dezember 2006, S. 18.

8 Jon Kessler a.a.O., Fn. 6.

9 In diesem Sinne noch Brian O'Doherty, Inside the White Cube. The Ideology of the Gallery Space (1976), deutsche Ausgabe Berlin 1996, S. 13 ff.

10 (2000), abgedruckt in: Gamma und Amplitude, Hamburg 2004, S. 320–344.

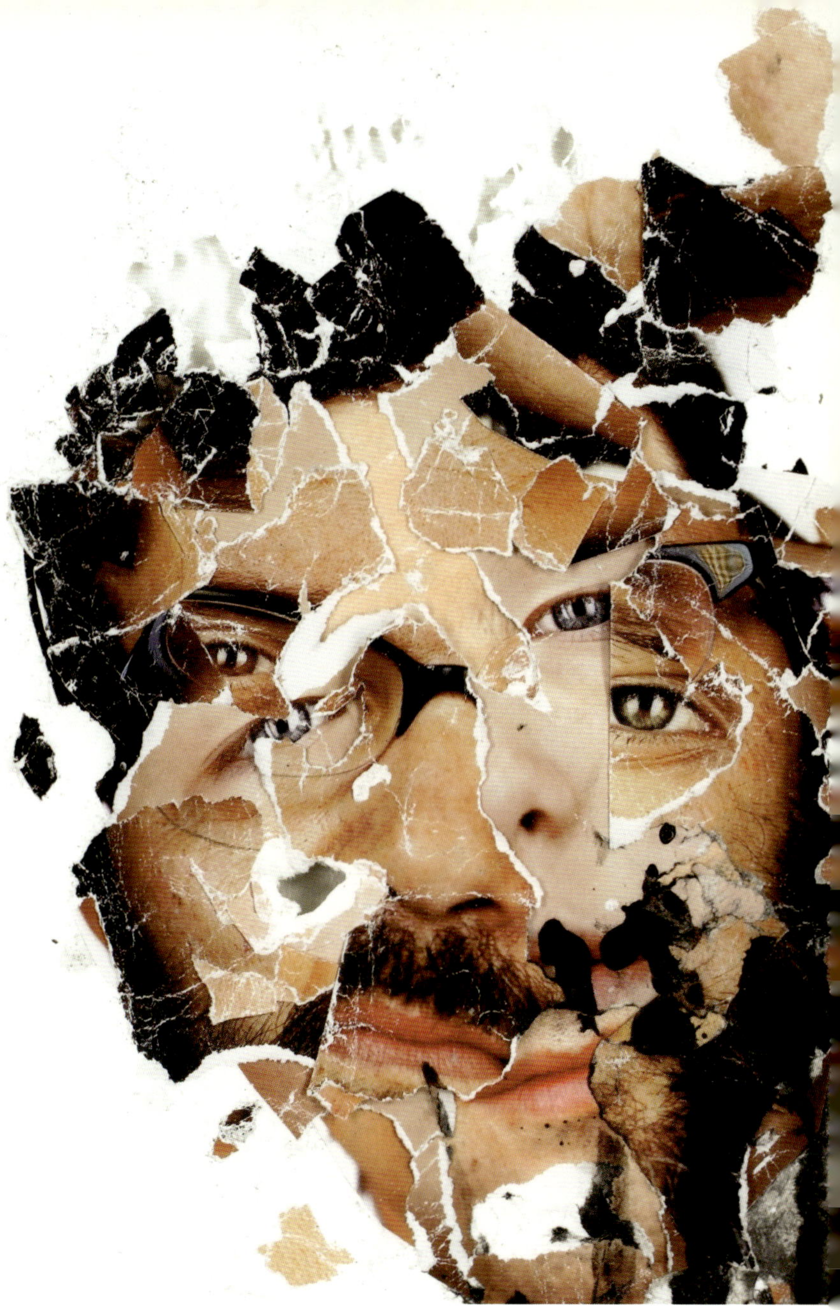

Works on Paper

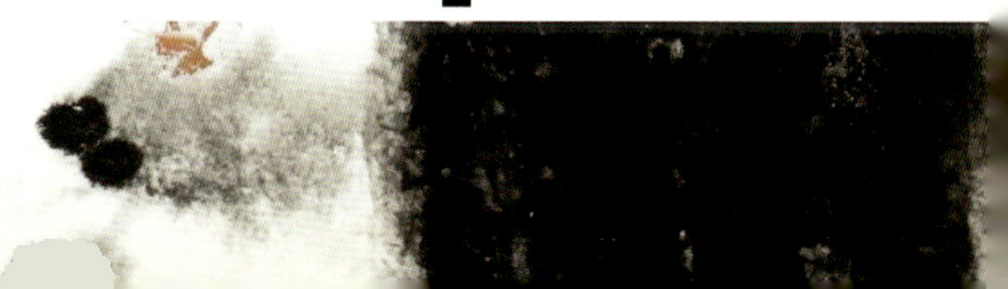

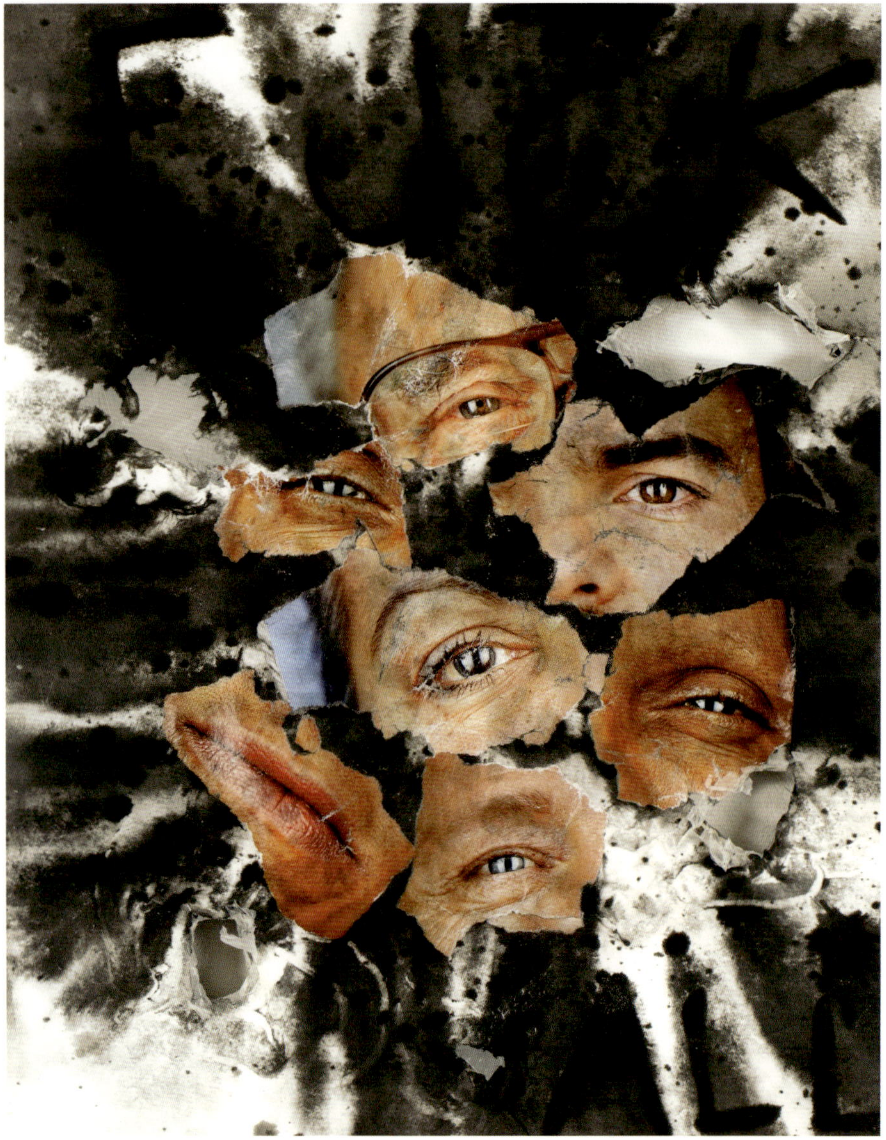

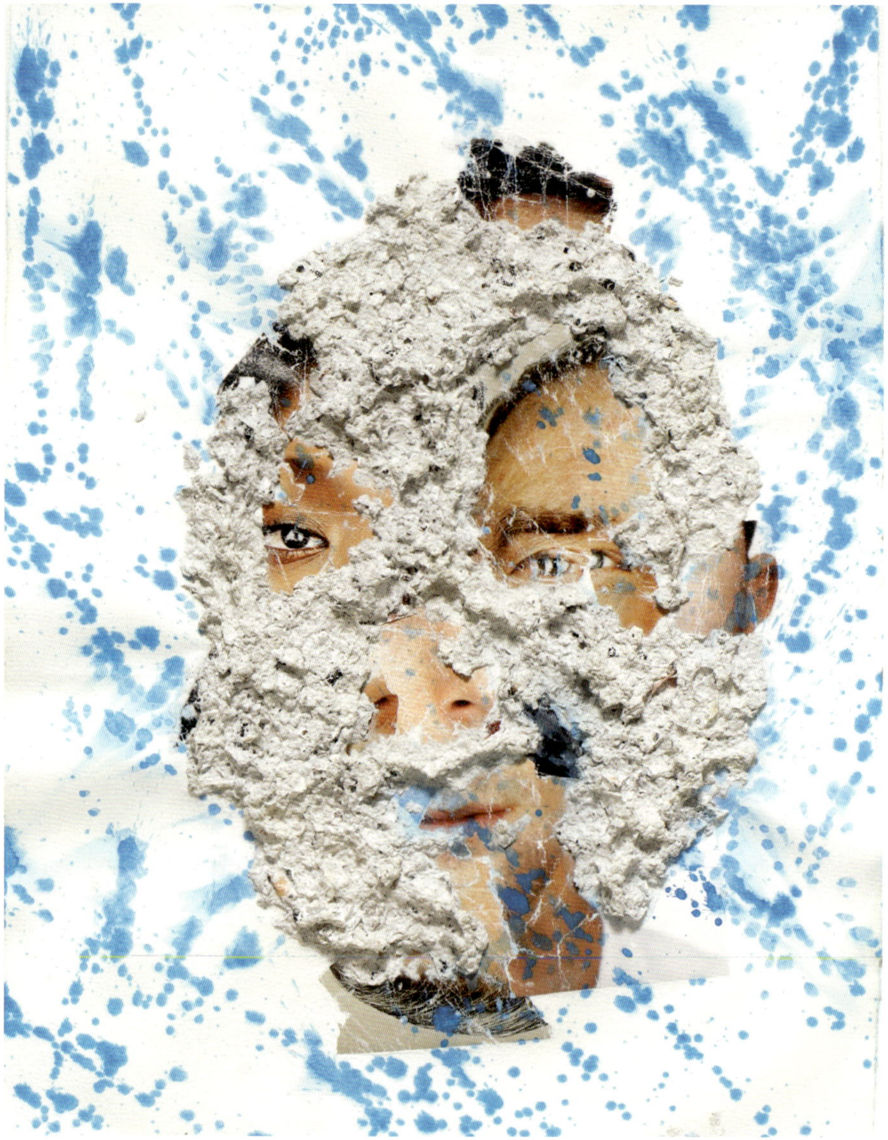

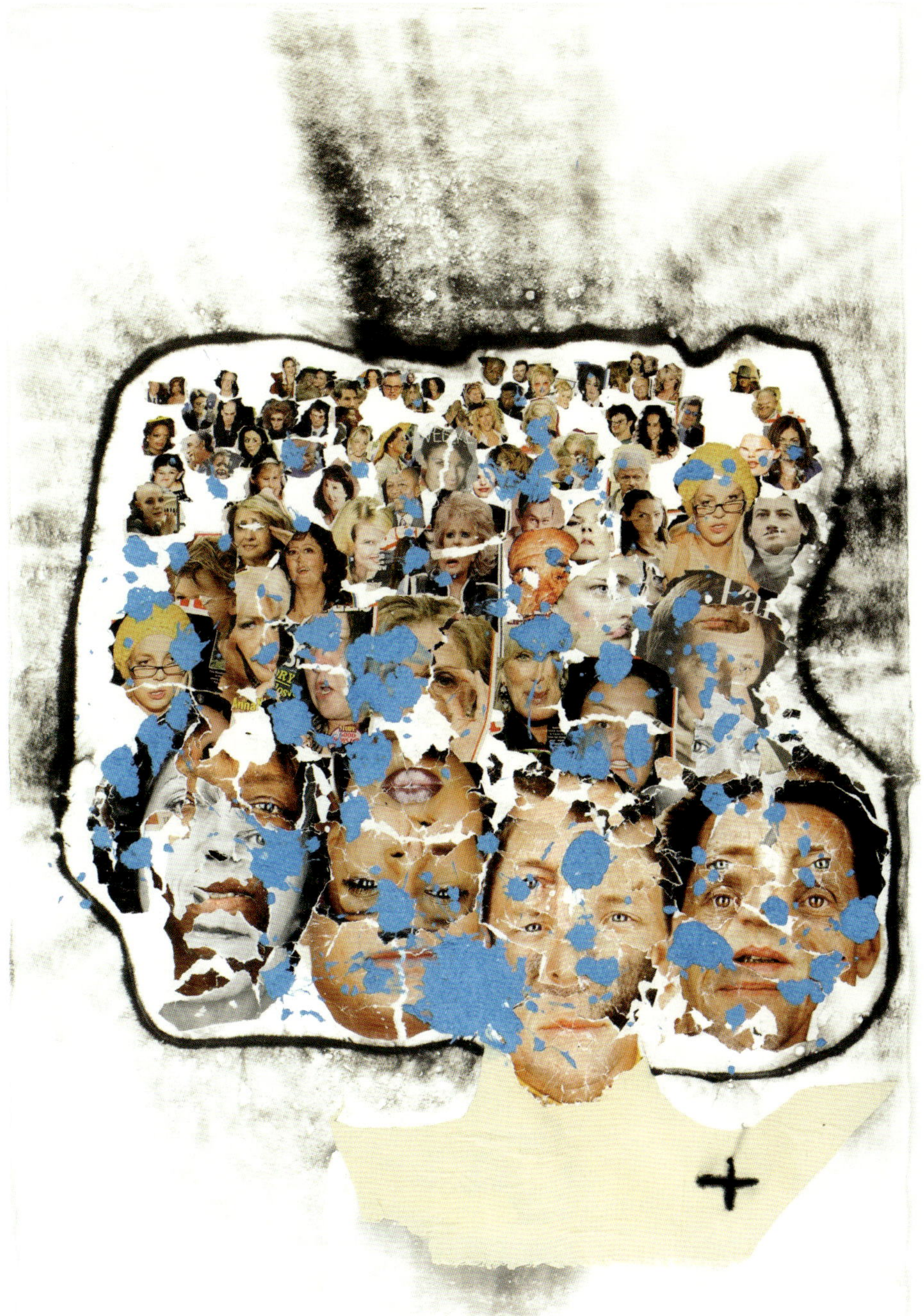

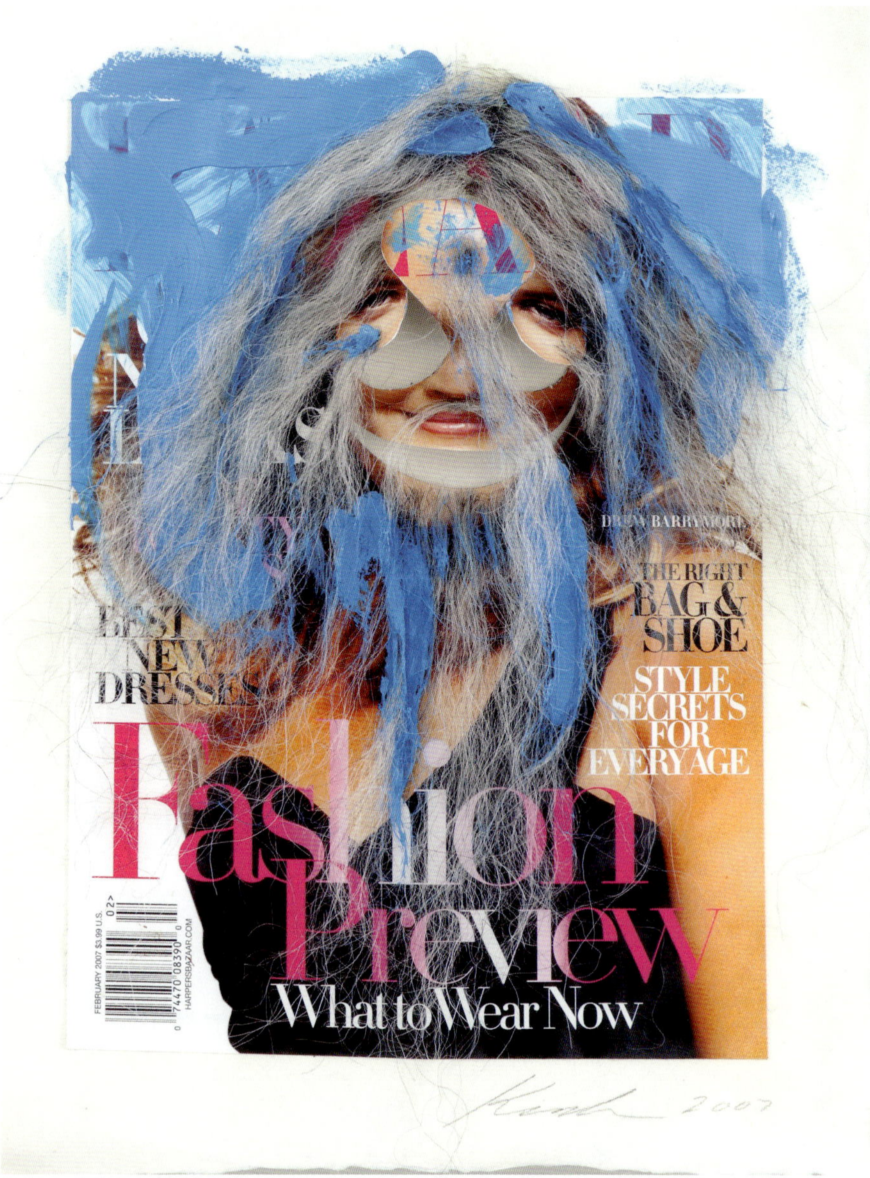
THE RIGHT
BAG &
SHOE
BEST
NEW
DRESSES
STYLE
SECRETS
FOR
EVERY AGE
Fashion
Preview
What to Wear Now
2007

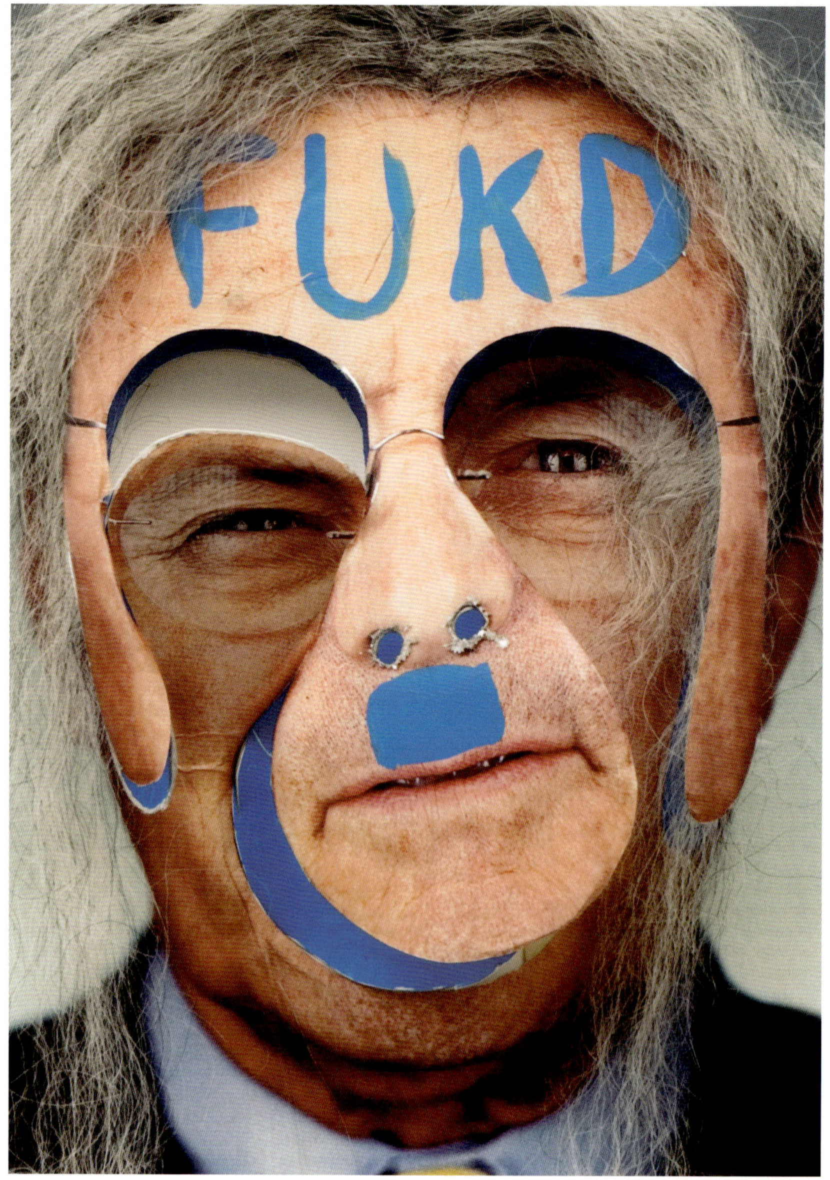
FUKD

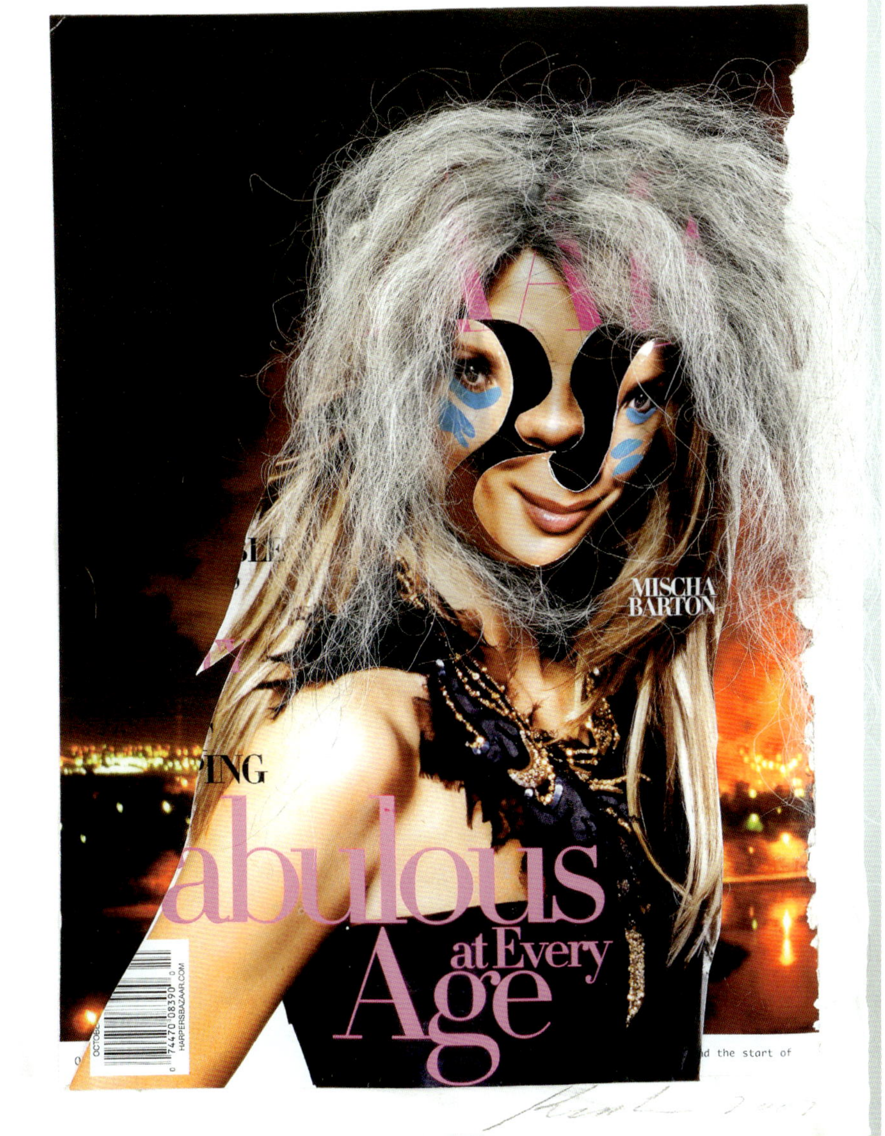
MISCHA
BARTON
ING
abulous
at Every
Age
HARPERSBAZAAR.COM
d the start of

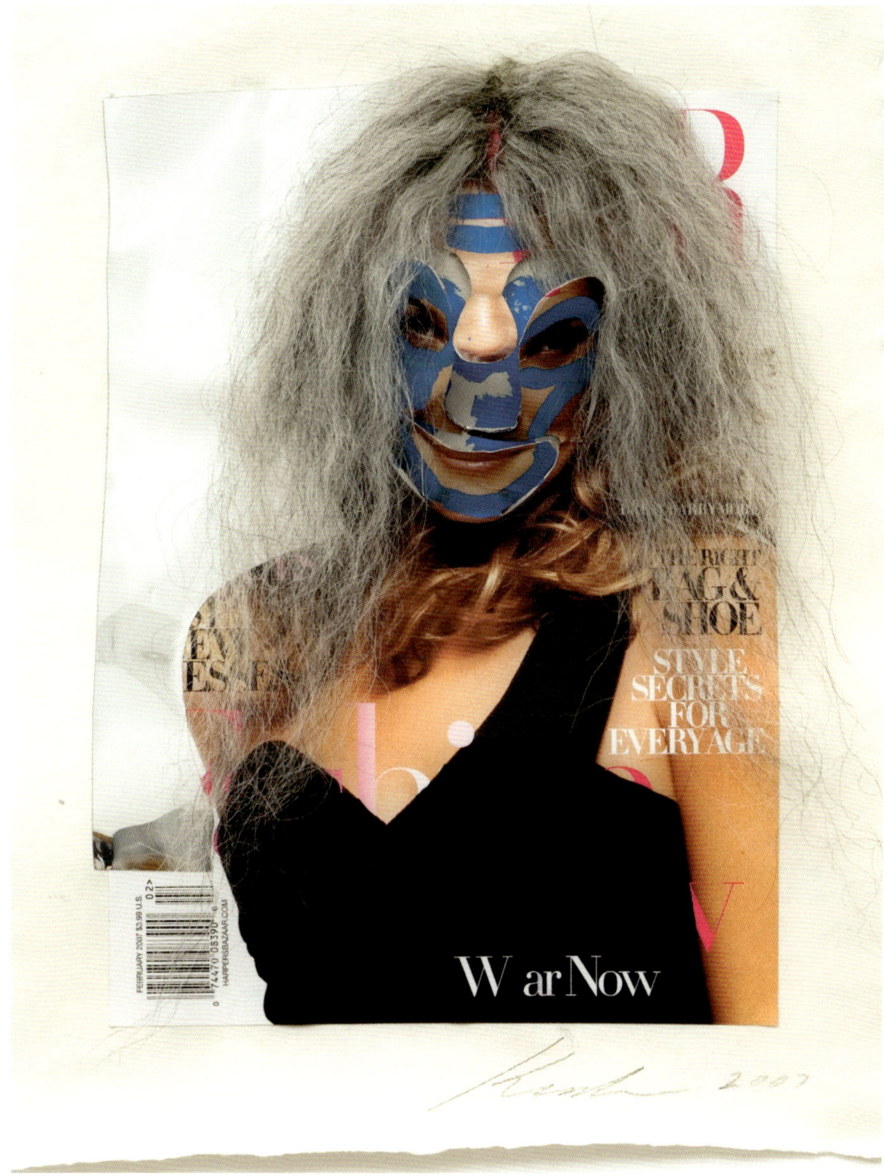
THE RIGHT
BAG &
SHOE
STYLE
SECRETS
FOR
EVERY AGE
W ar Now
HARPERSBAZAAR.COM

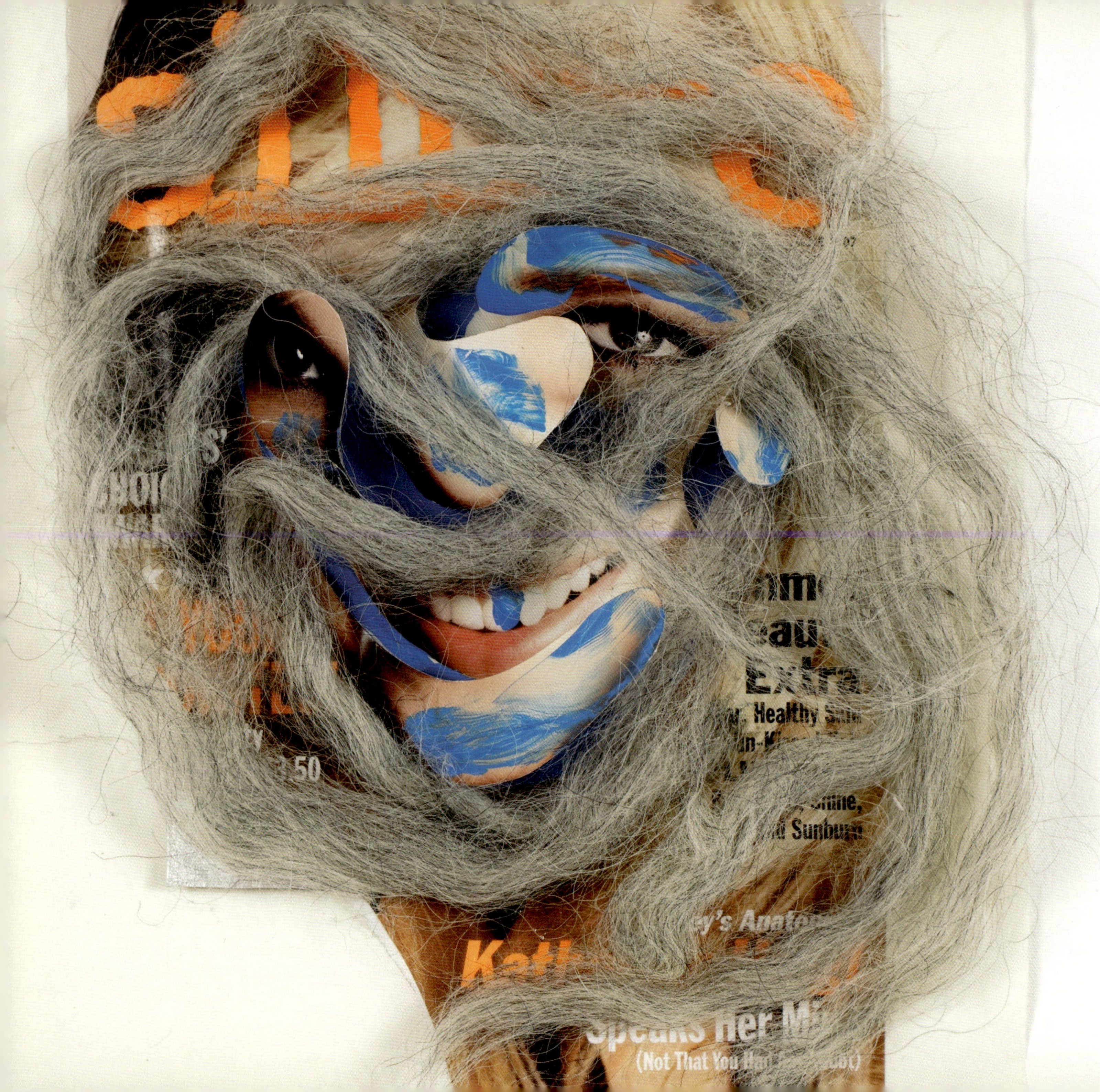

Extra
Healthy
Sunburn
Speaks Her
(Not That You Had

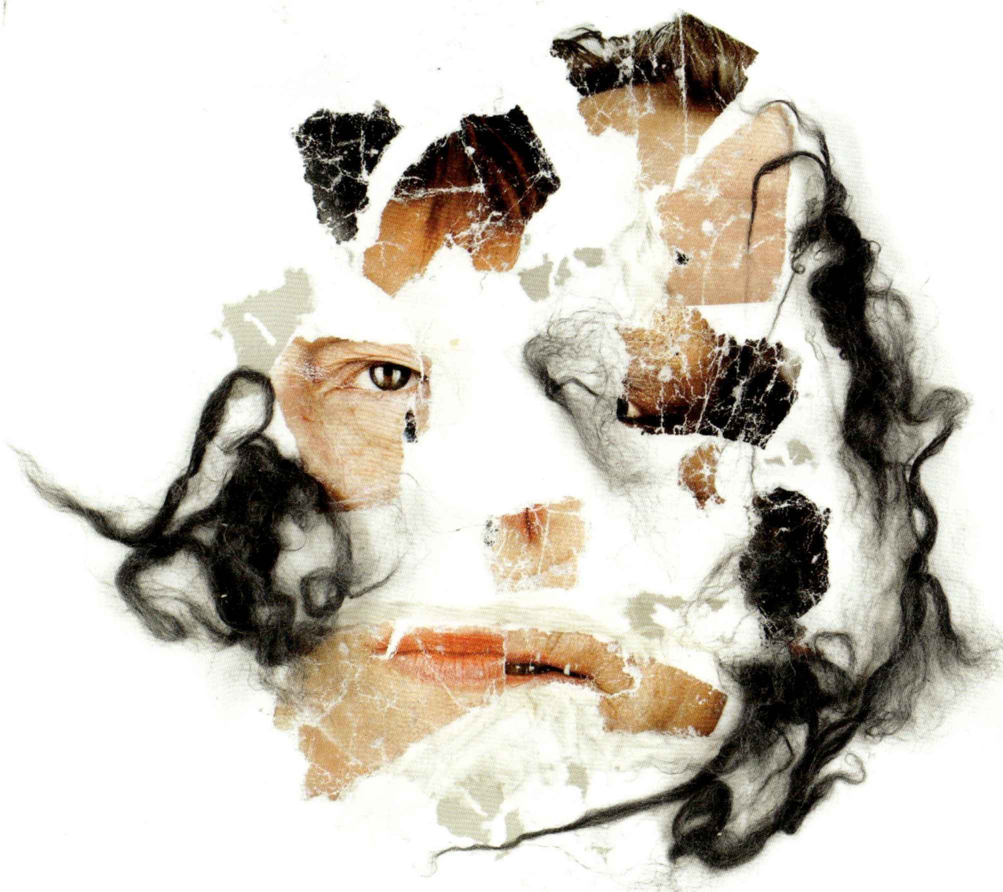

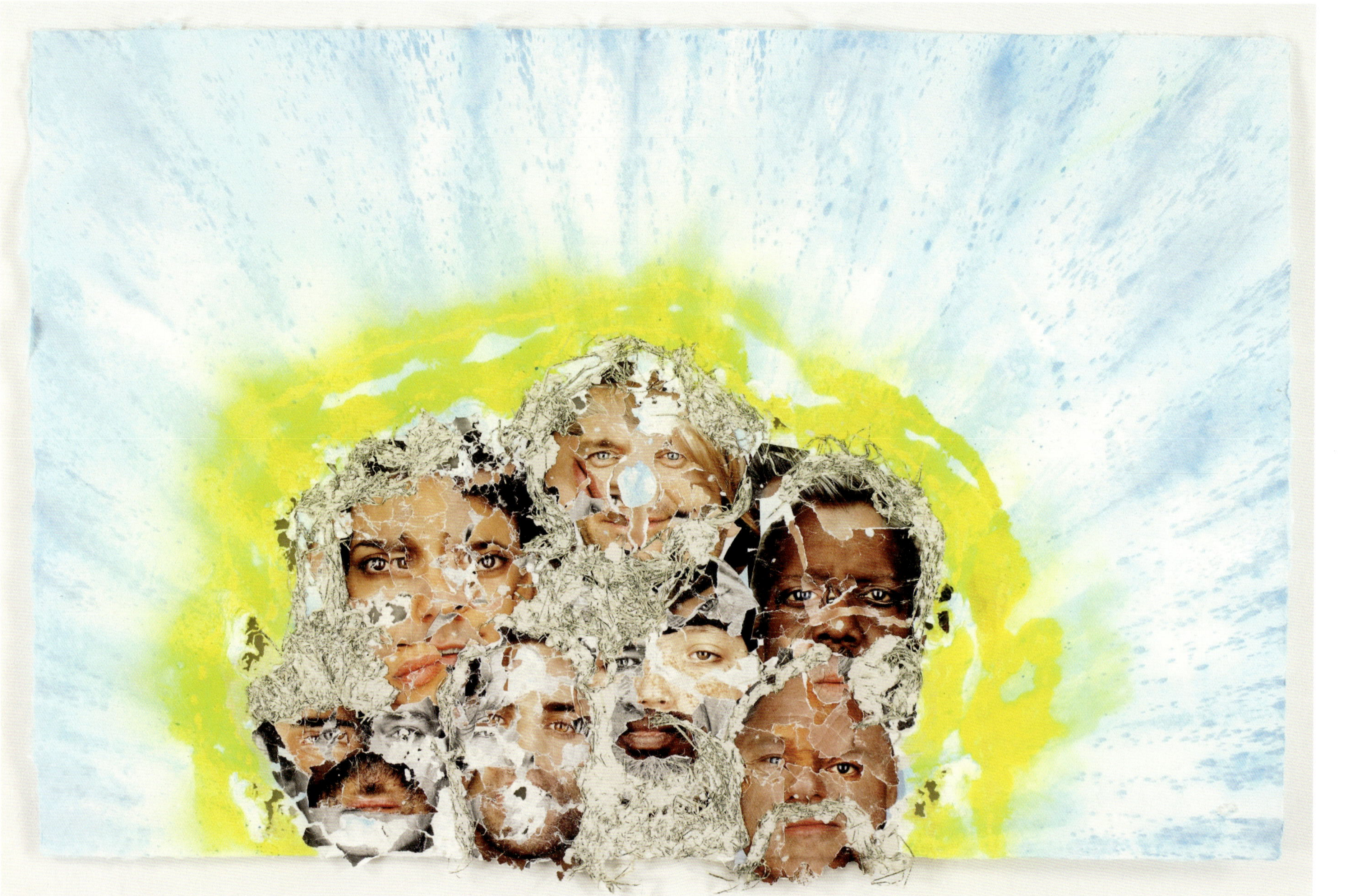

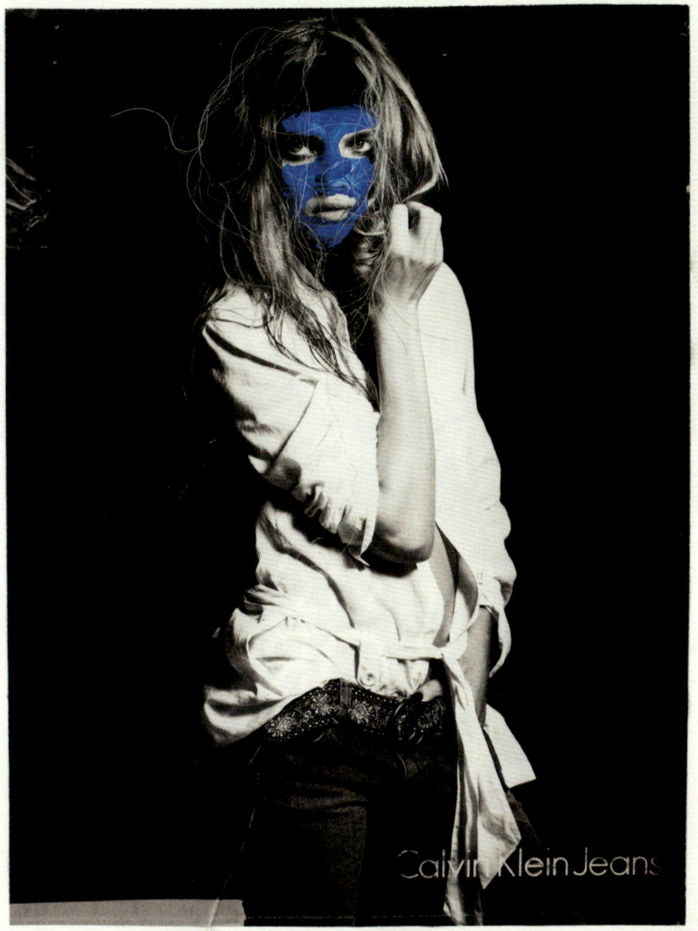
Calvin Klein Jeans

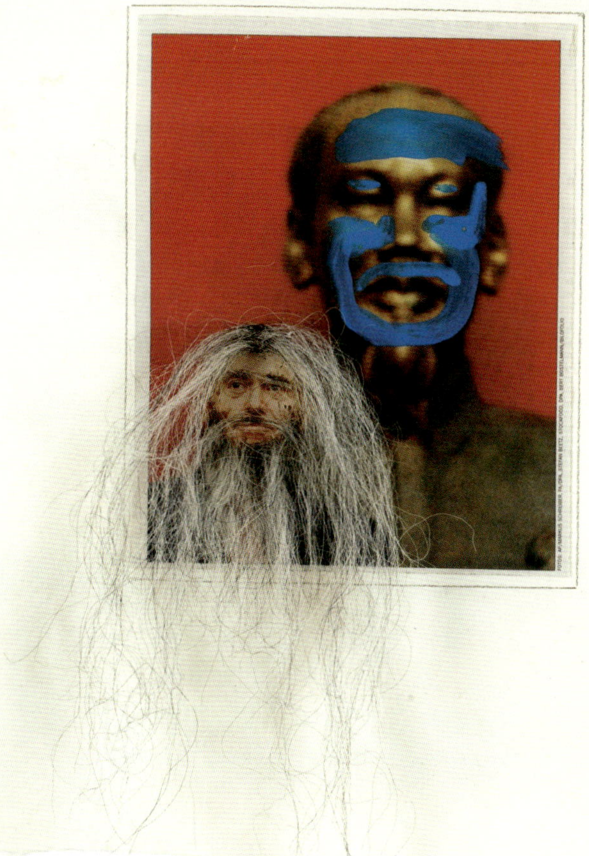

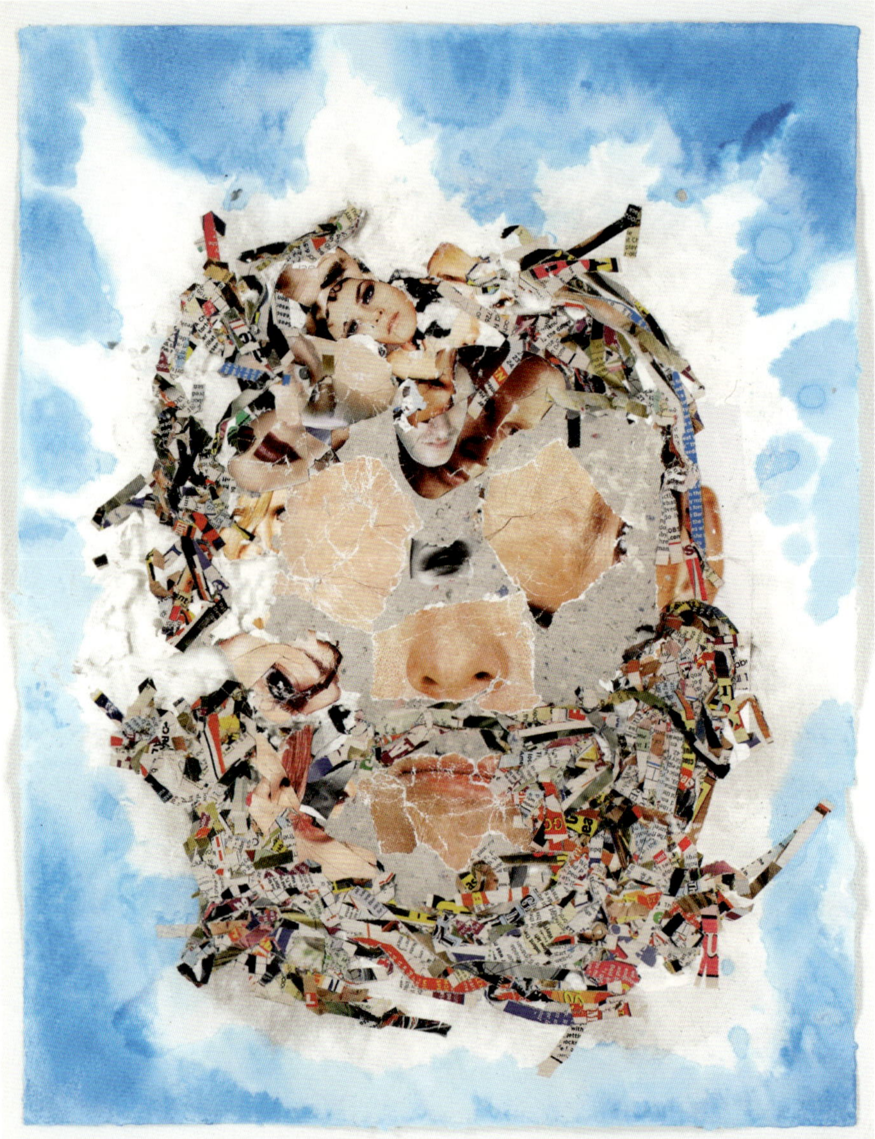

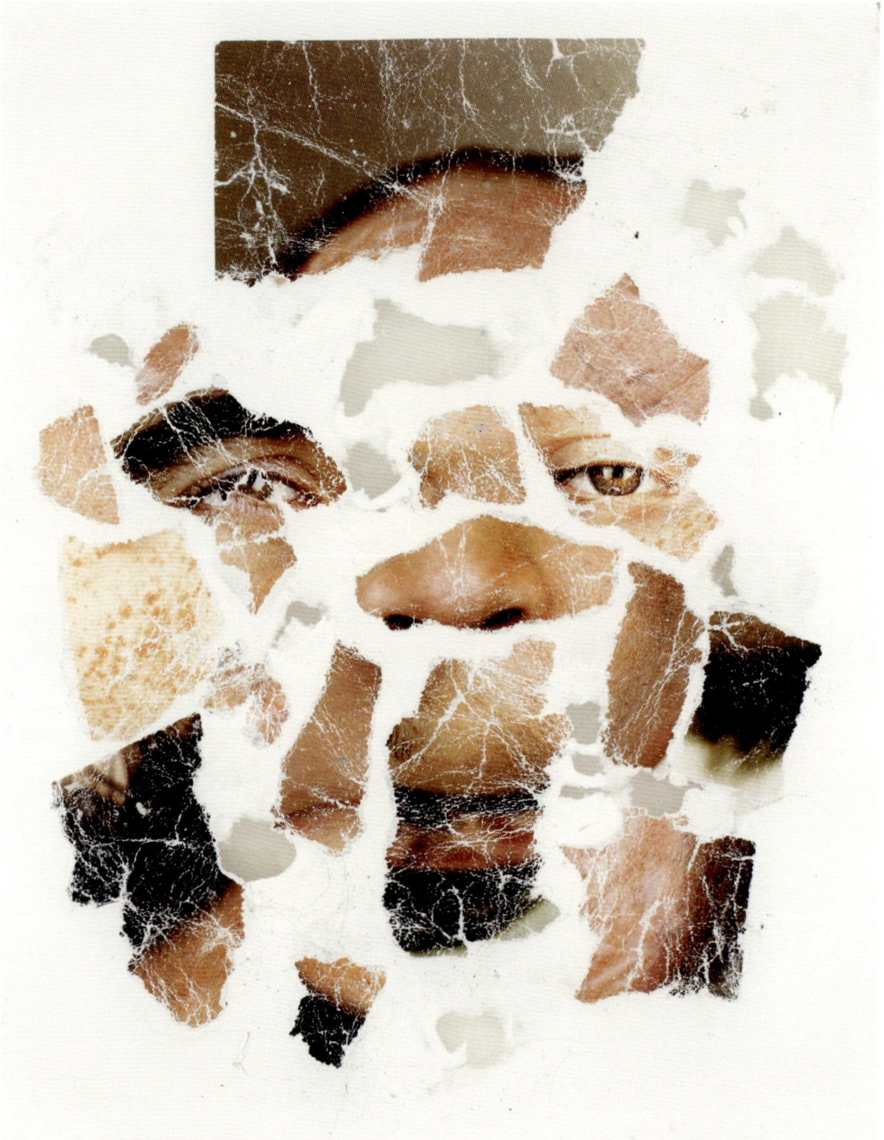

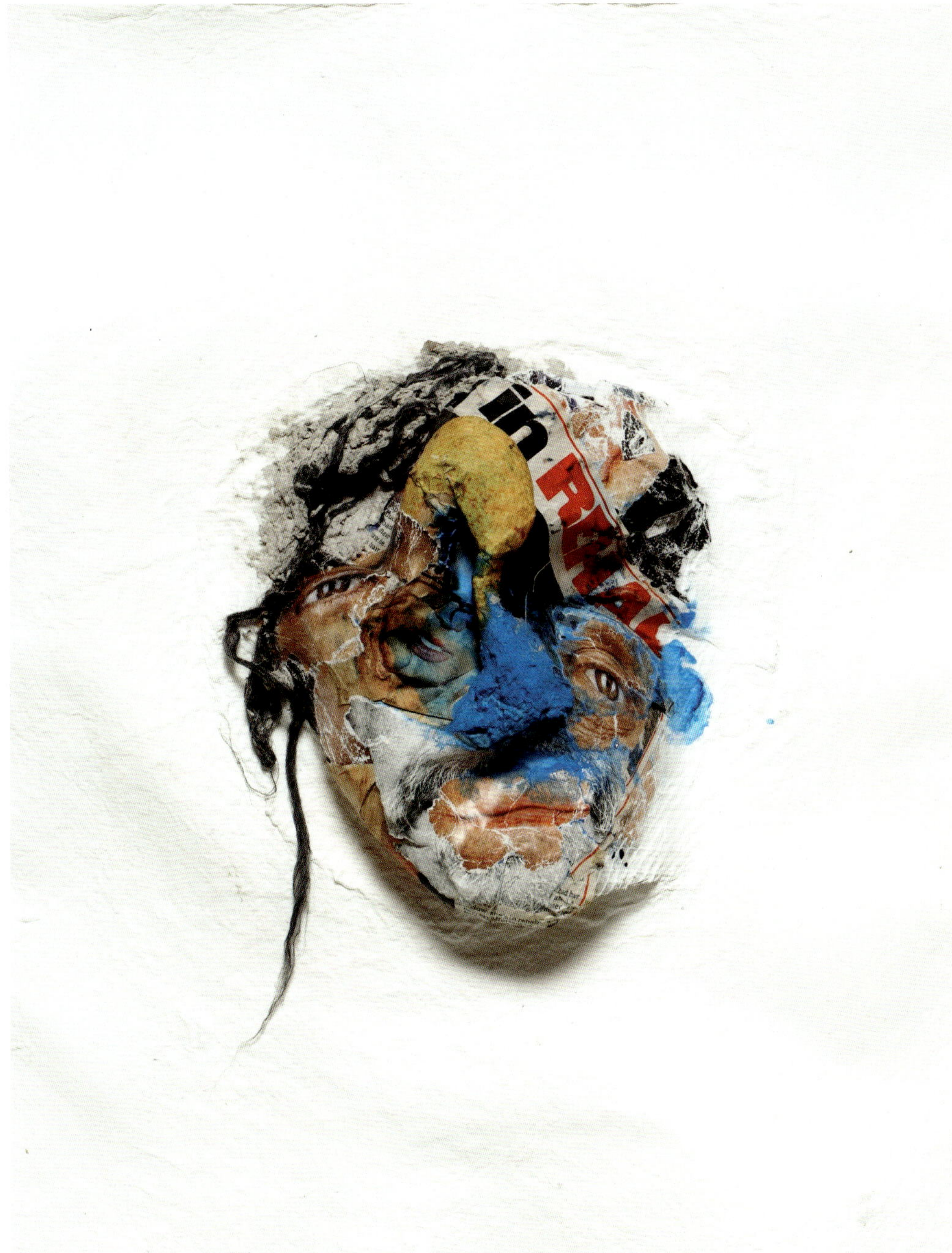

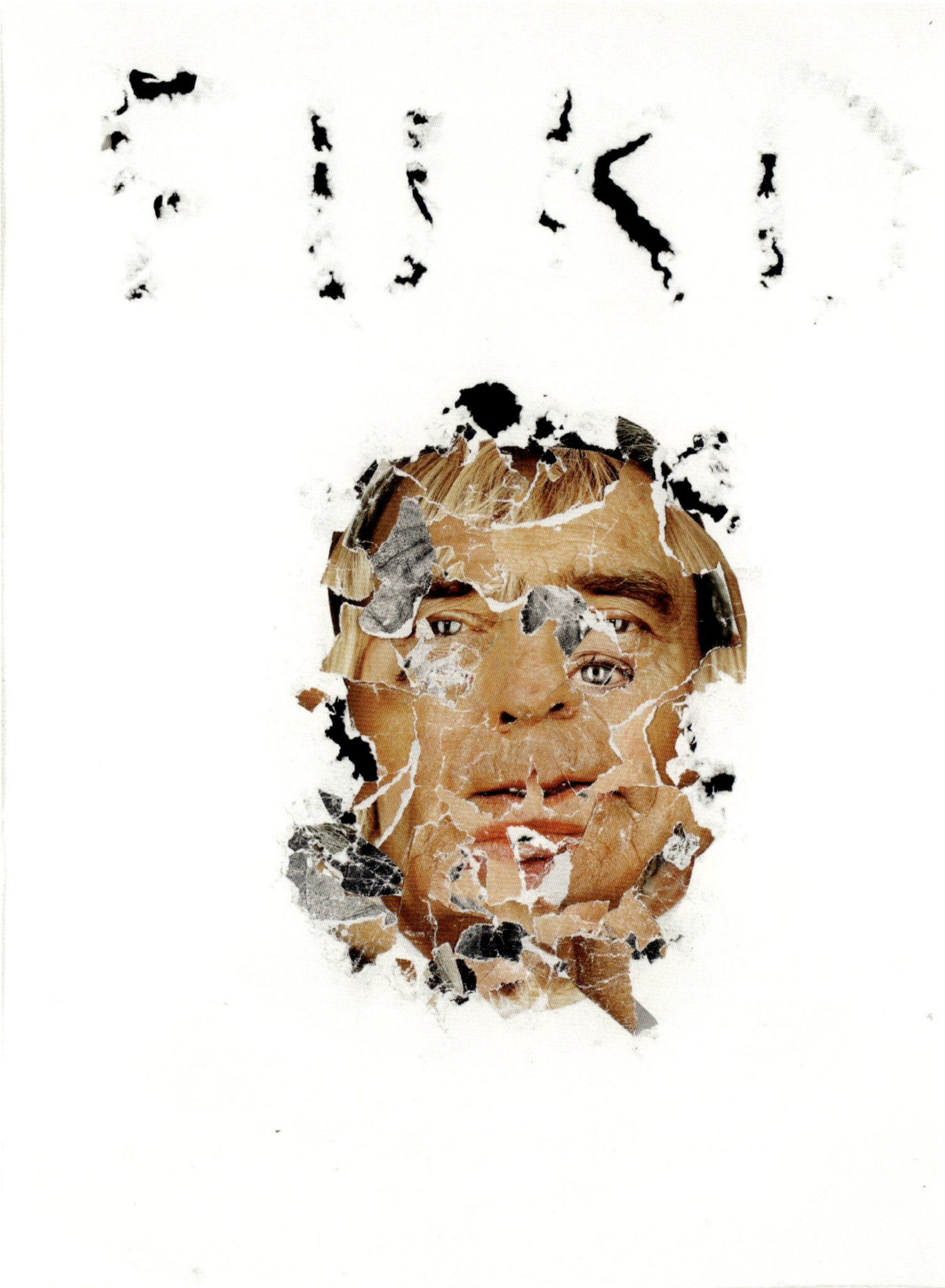

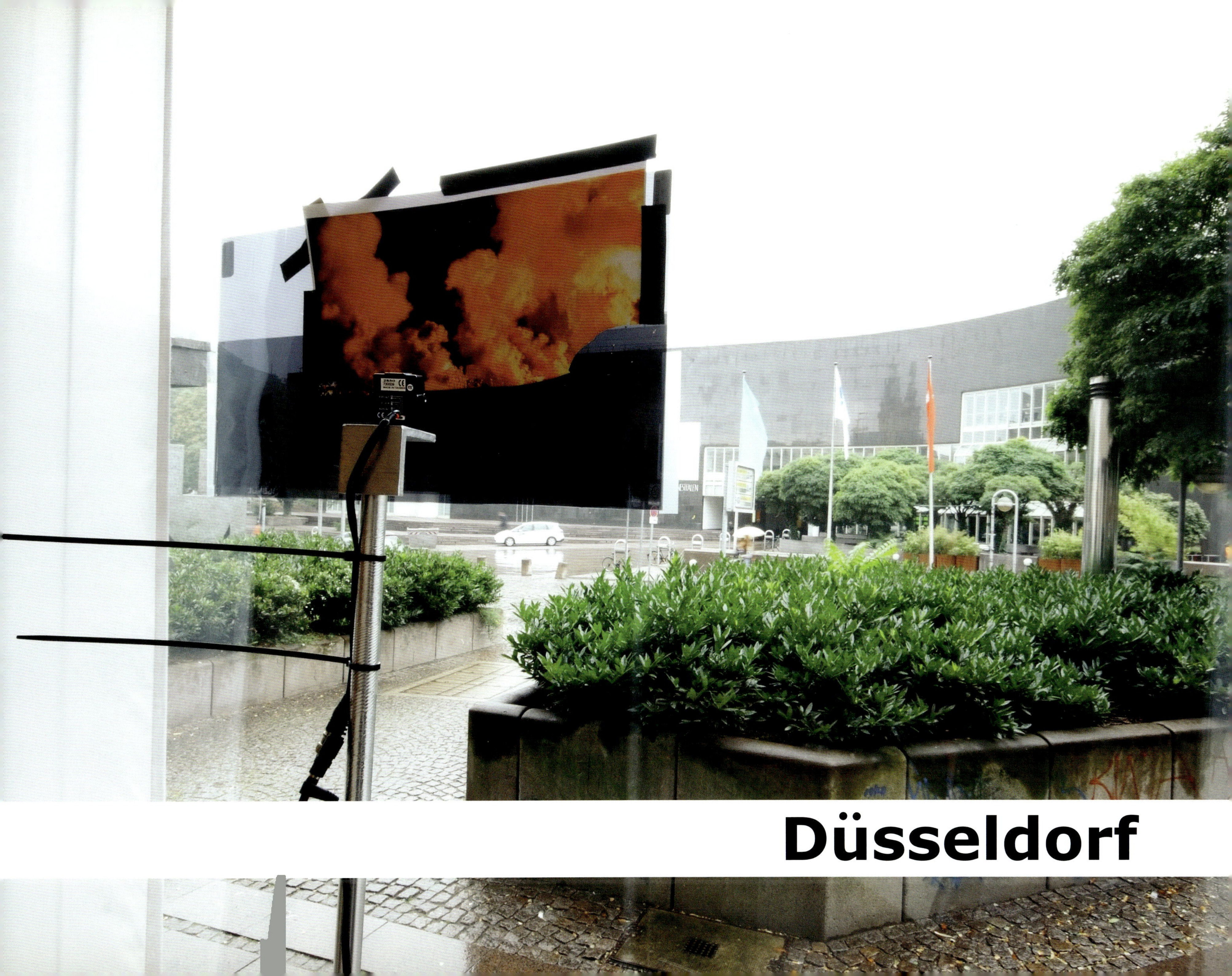

Düsseldorf

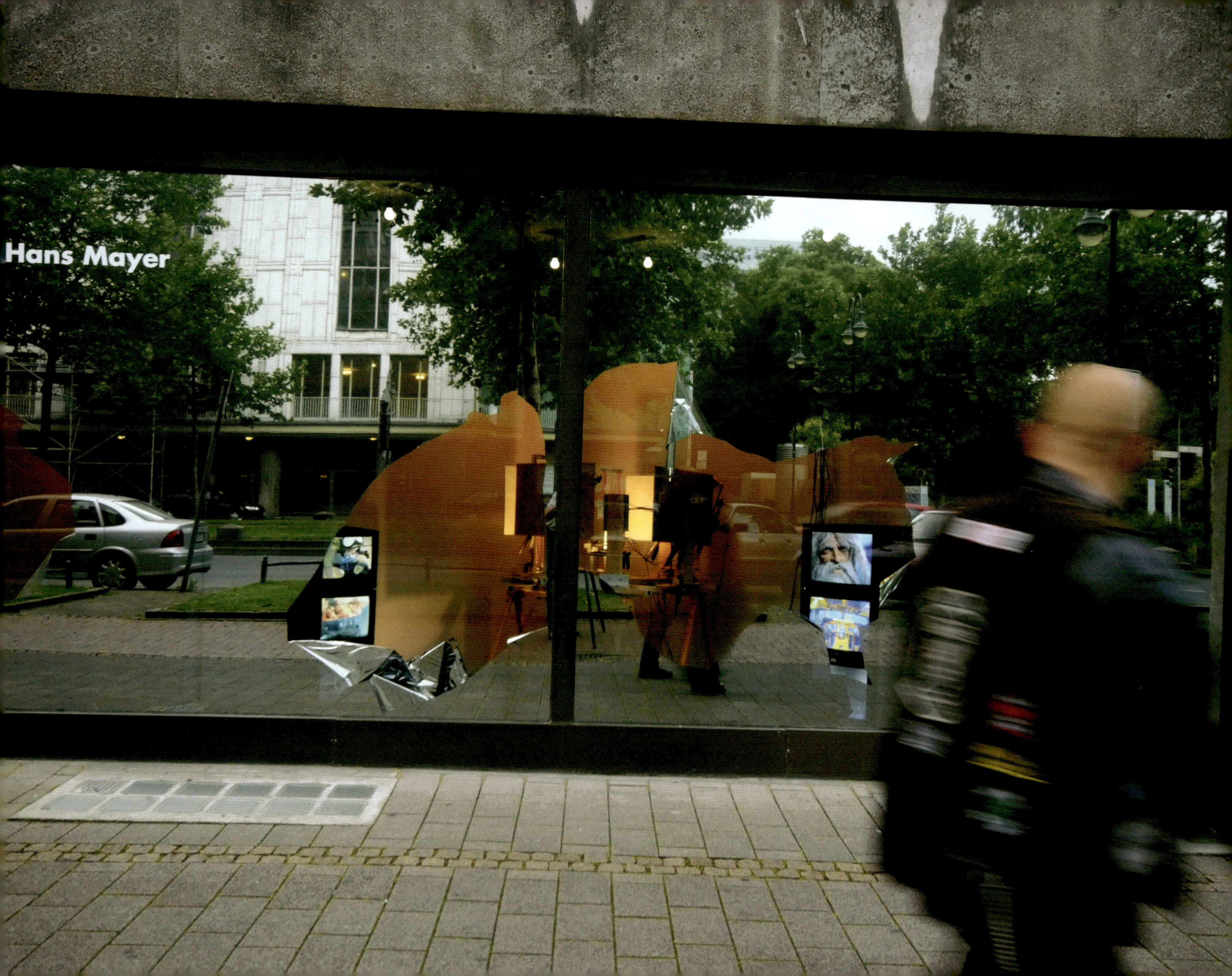
Hans Mayer

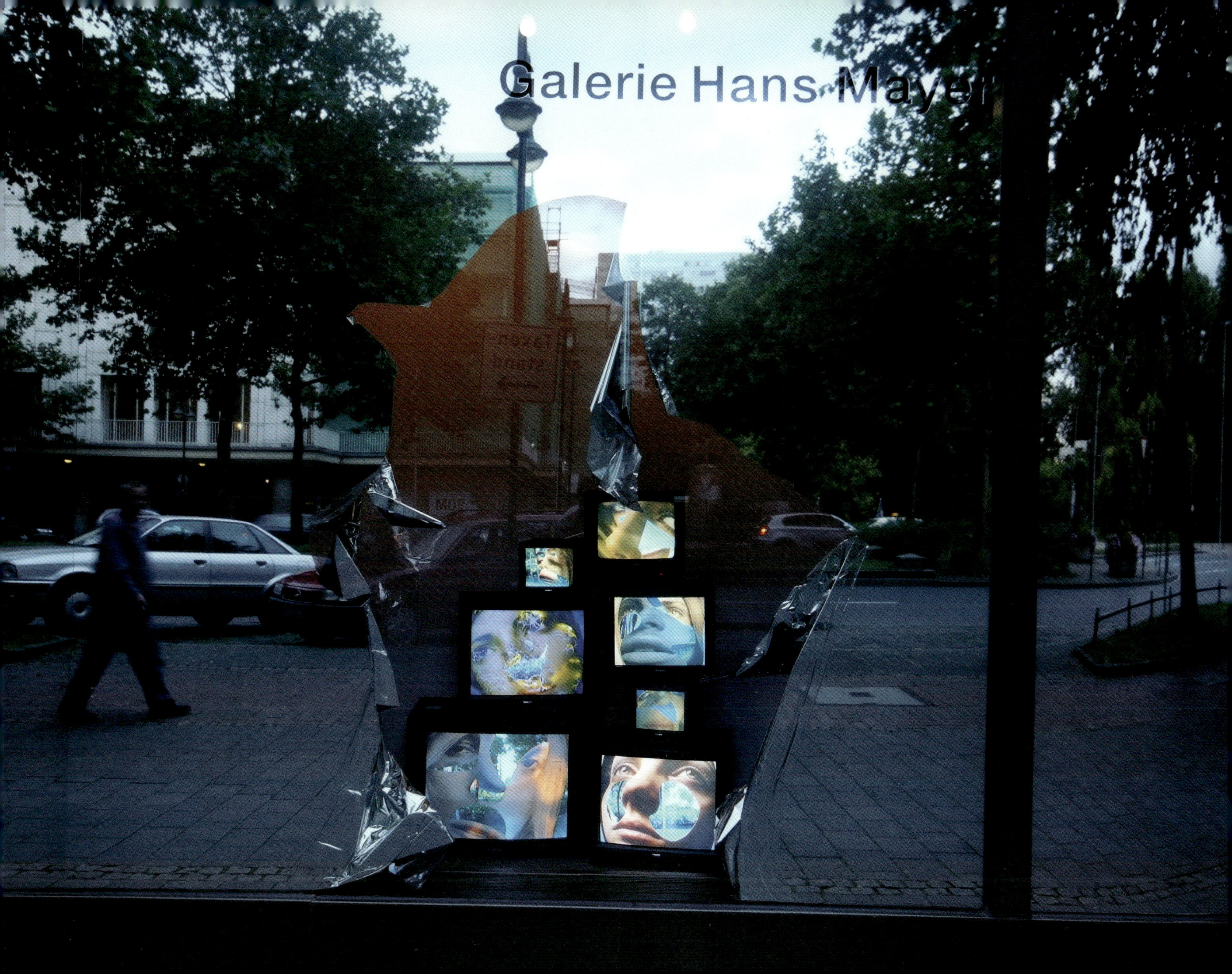
Galerie Hans Mayer

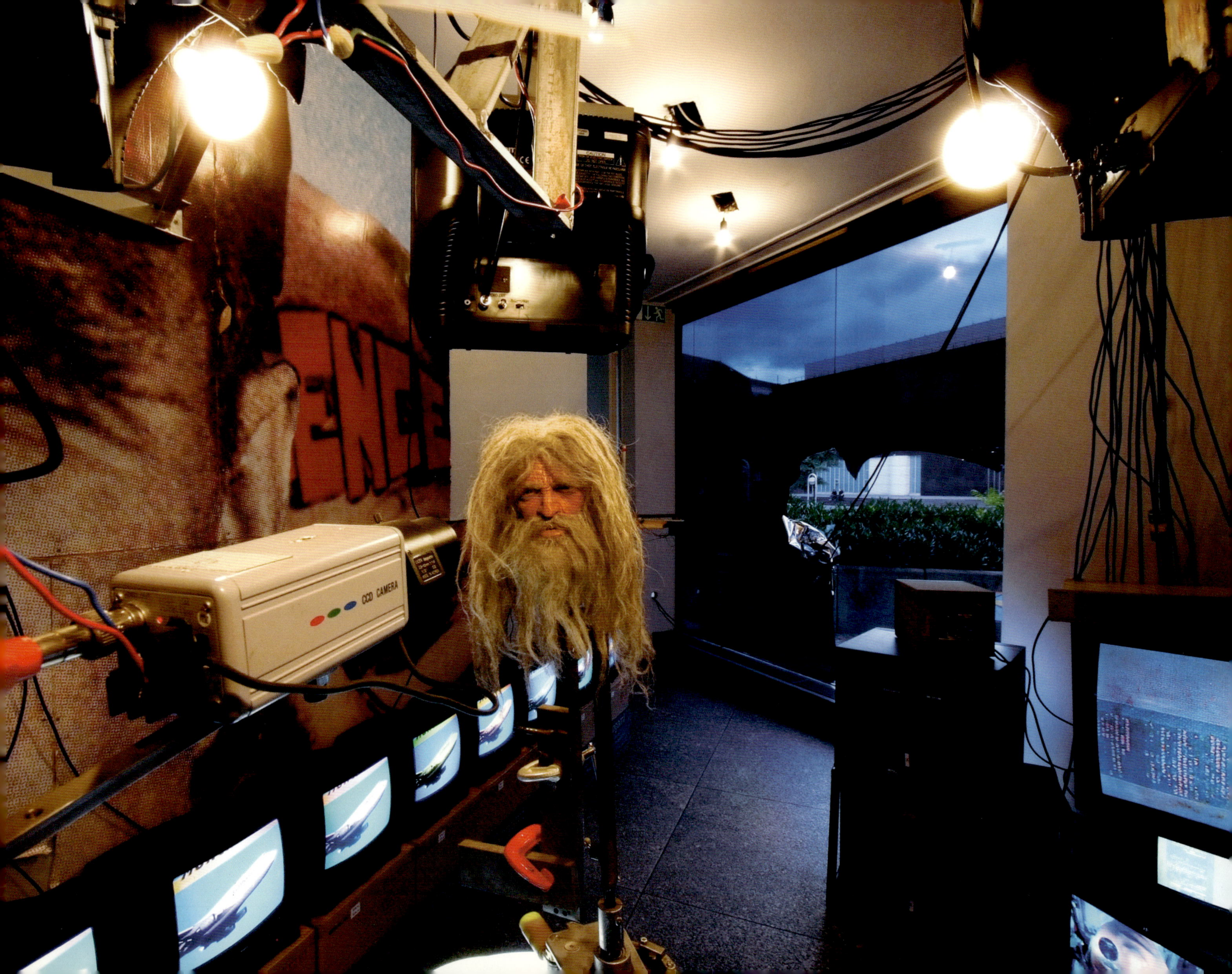
CCD CAMERA

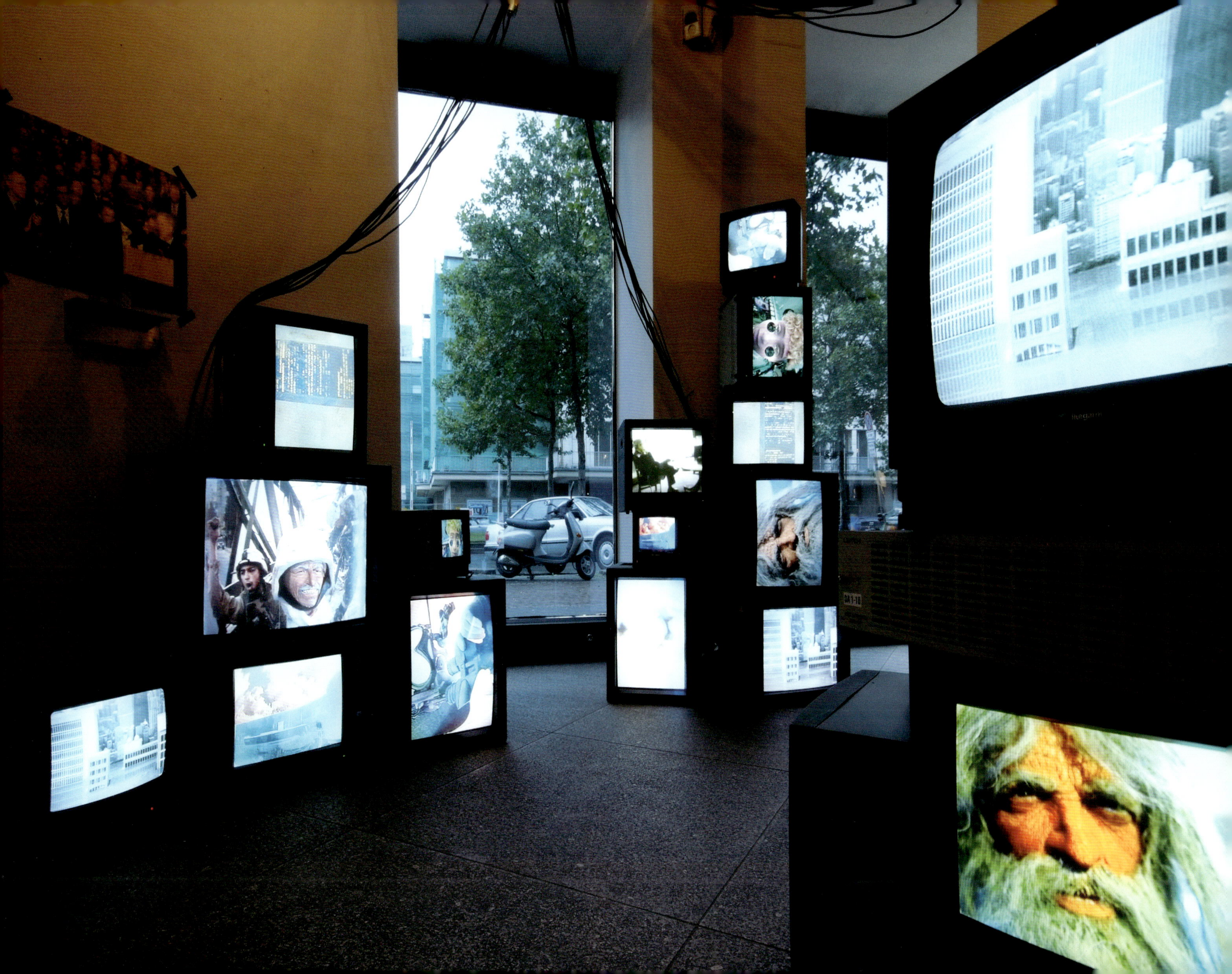

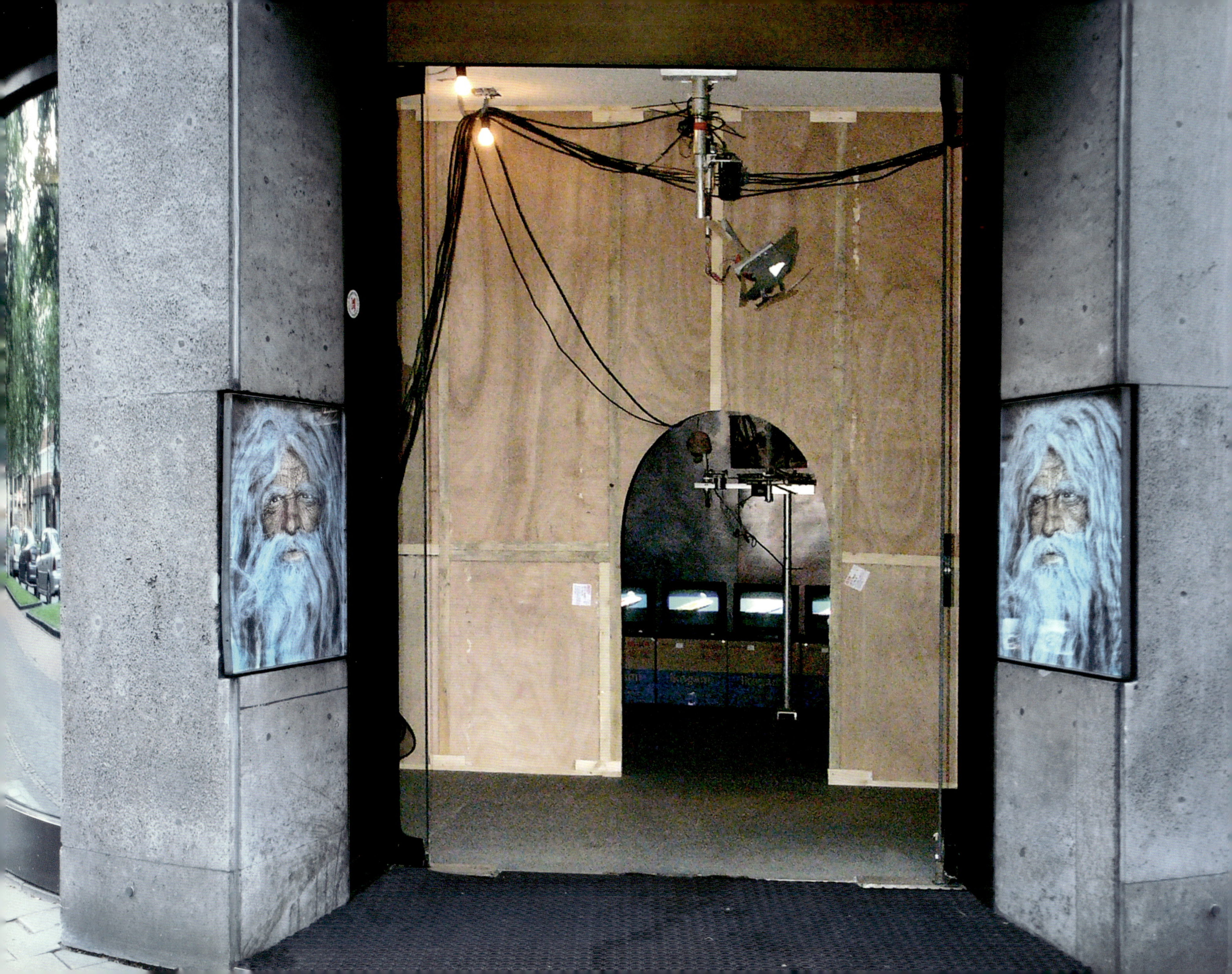

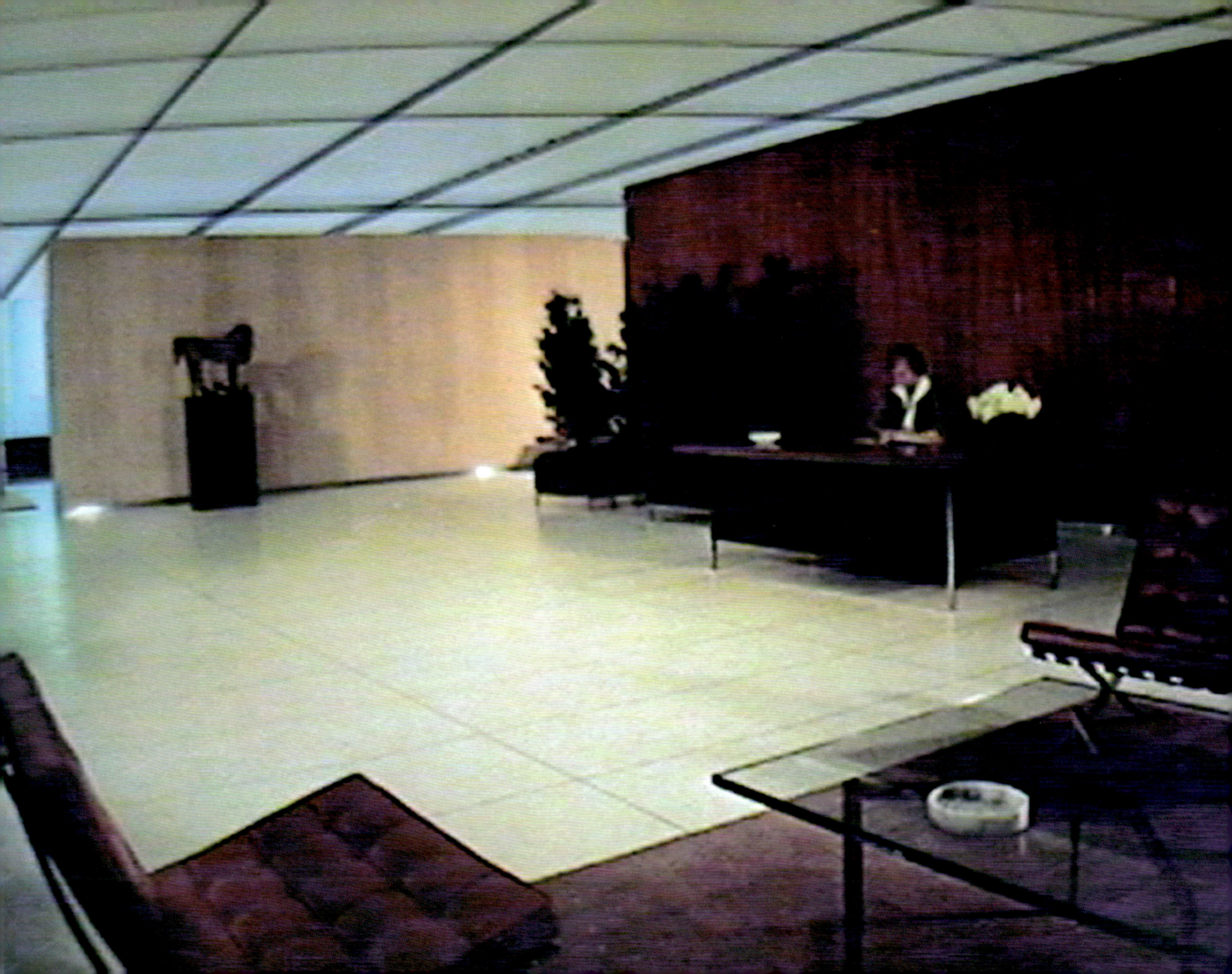

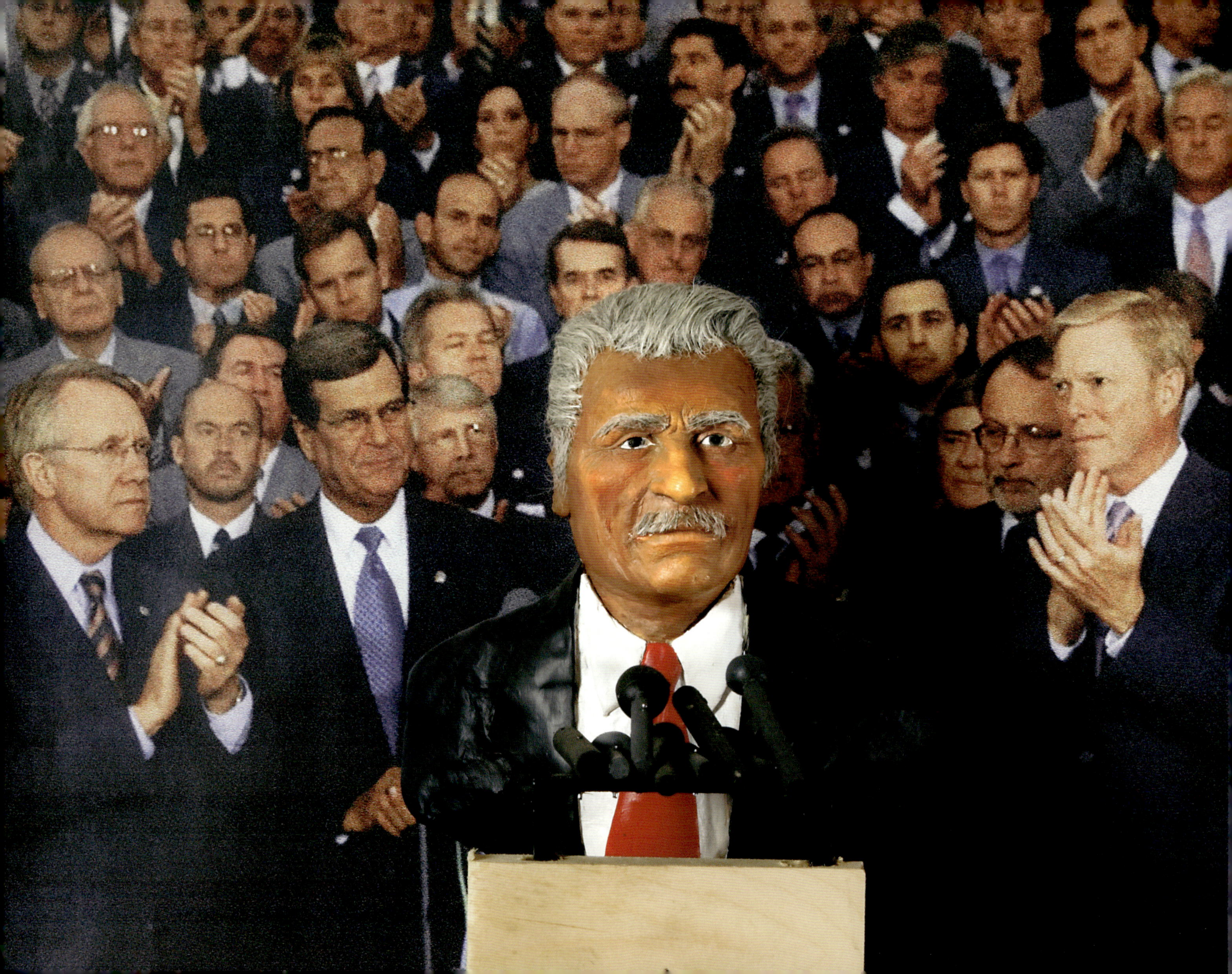

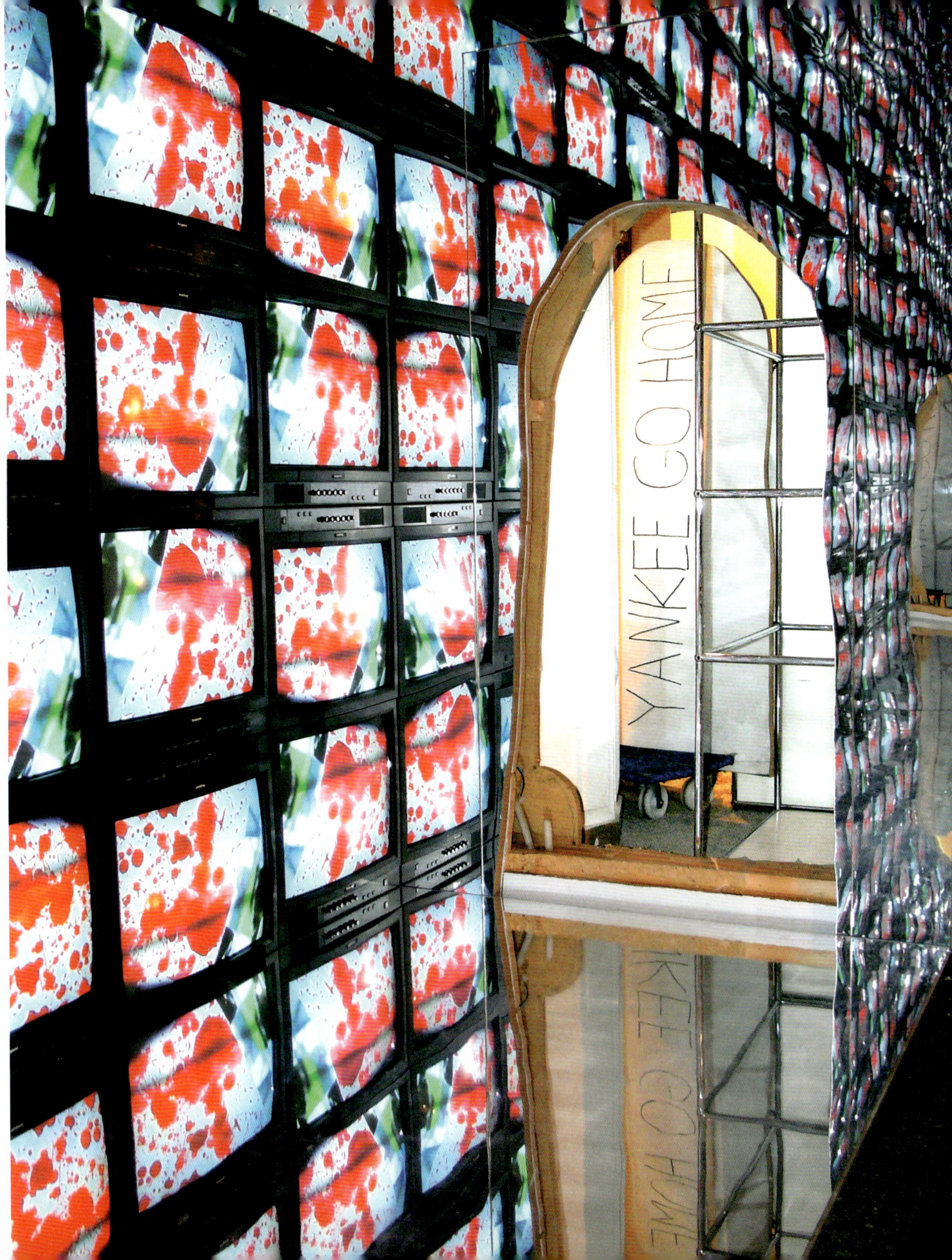
YANKEE GO HOME

Biography

Jon Kessler was born in Yonkers, New York, in 1957. He graduated with a BFA from SUNY College at Purchase and attended the Whitney Museum Independent Study Program.
His first exhibition was at Artist's Space in 1983, followed by a show at White Columns six months later.
Since that time he has exhibited widely in galleries and museums in the United States, Europe, and Japan. Kessler is a recipient of the Guggenheim Fellowship, the Foundation for the Performing Arts Grant, the St. Gaudens Sculpture Award, and several grants from the National Endowment for the Arts.
His sculptures can be found in many private and public collections, including The Museum of Modern Art, The Whitney Museum of American Art, and MOCA, Los Angeles. He is a professor of Visual Arts at Columbia University and lives in New York City.

Biographie

Jon Kessler ist in Yonkers, New York, in 1957 geboren. Er diplomierte in Bildender Kunst in dem *SUNY College at Purchase* und besuchte das *Whitney Museum Independent Study Program*. Seine erste Ausstellung fand in den *Artist's Space* in 1983 statt. Sechs Monate später wurde eine Darbietung bei den *White Columns* vorgeführt. Seitdem hat er in vielen verschiedenen Galerien in den Vereinigten Staaten von Amerika, in Europa und in Japan ausgestellt. Kessler bekommt die *Guggenheim Mitgliedschaft*, das Stipendium von der *Foundation for the Performing Arts*, der *St. Gaudens Sculpture Award*, und mehrere Stipendiums von der Nationalen Stiftung für die Künste.
Seine Skulpturen befinden sich in vielen privaten und öffentlichen Sammlungen, wie in *The Museum of Modern Art*, *The Whitney Museum of American Art*, und *MOCA*, Los Angeles. Er ist Professor von Bildender Kunst in der Columbia University und lebt in New York.

The artist would like to thank

Asti Hustvedt
Harald Falckenberg
Alanna Heiss
Hans Mayer
Jeffrey Deitch
Larissa Hilbig
Brett Littman
Nate Longcope
Ohad Meromi
Robert Longo
Antoine Guerrero
Tom Powel
Rick Haylor
Hal Foster
Ludwig Seyfarth
Martin Basher
Douglas Irving Repetto
Tonia Schultz
Jesse Finkelstein
Hartmut Gerbsch
Steven Henry Madoff
Isabelle Woodley
Art Agents
Yng-Ru Chen
Peter Russo
Rachel Gladfelter
Dieu Donne Papermill
Andrew Ohanesian
Volker Trautner
Richard Lohr
Sam Frank
Andrea Cashman
Martje Schulz

Contents

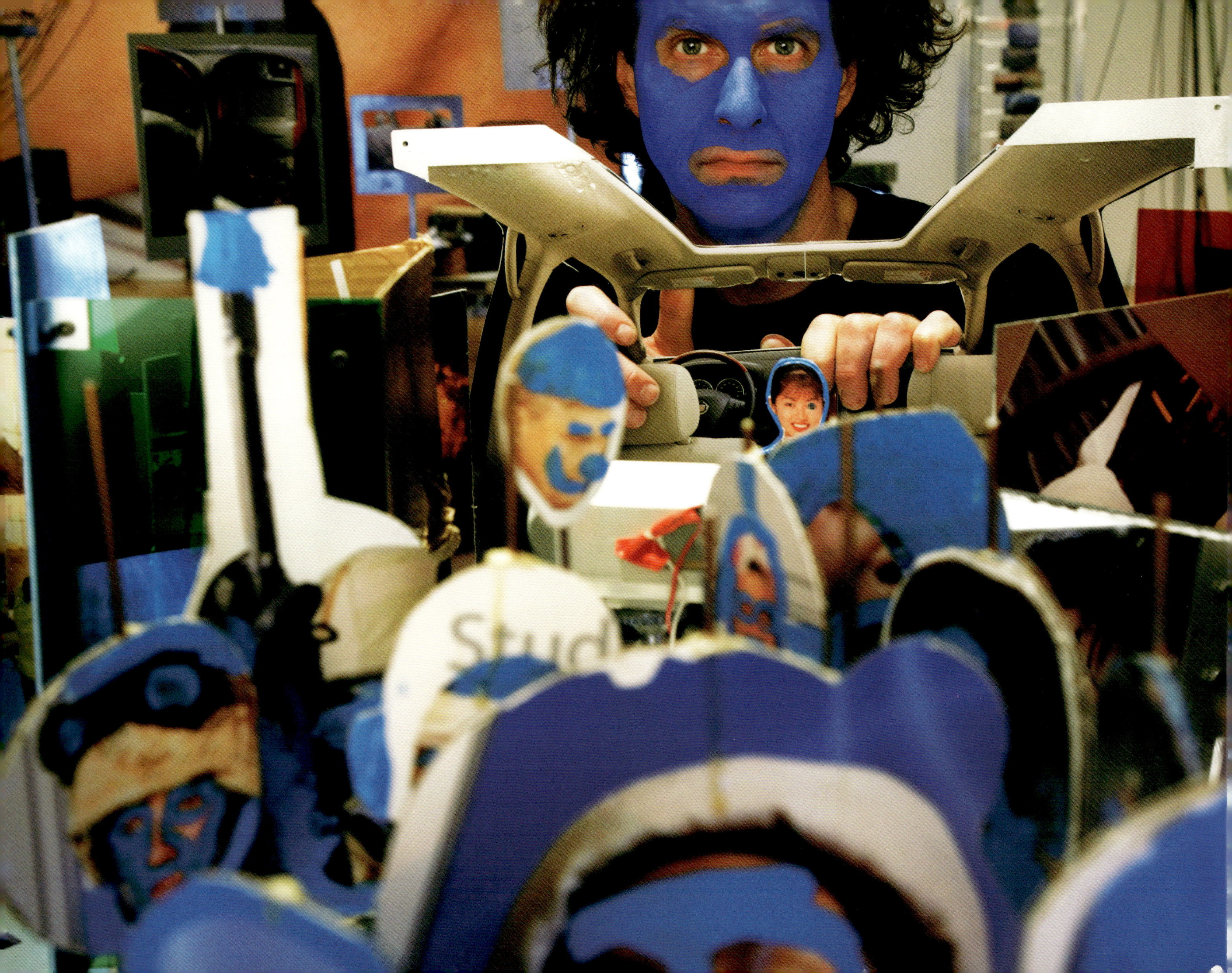